Geoheritage, Geoparks and Geotourism

Conservation and Management Series

Spectacular geo-morphological landscapes and regions with special geological features or mining sites, are becoming increasingly recognized as critical areas to protect and conserve for the unique geoscientific aspects they represent and as places to enjoy and learn about the science and history of our planet. More and more national and international stakeholders are engaged in projects related to "Geoheritage", "Geo-conservation", "Geoparks" and "Geo-tourism" and are positively influencing the general perception of modern Earth sciences. Most notably, "Geoparks", have proven to be excellent tools to educate the public about "Earth Sciences". And shown to be areas for recreation and significant sustainable economic development through geotourism. In order to develop further the understanding of earth sciences in general and to elucidate the importance of earth sciences for Society the Geoheritage, Geoparks and Geotourism Conservation and Management Series has been launched together with its sister GeoGuides series. "Projects" developed in partnership with UNESCO, World Heritage and Global Geoparks Networks, IUGS and IGU, as well as with the "Earth Science Matters" Foundation, are welcome. The series aims to provide a place for in-depth presentations of developmental and management issues related to Geoheritage and Geotourism as well existing and potential Geoparks. Individually authored monographs as well as edited volumes and conference proceedings are welcome in this series. This book series is considered to be complementary to the Springer-Journal "Geoheritage".

More information about this series at http://www.springer.com/series/11639

Daniel Kelley • Kevin Page
Diego Quiroga • Raul Salazar

In the Footsteps of Darwin: Geoheritage, Geotourism and Conservation in the Galapagos Islands

Springer

Daniel Kelley
School of Natural Resources
Hocking College
Nelsonville, OH, USA

Kevin Page
Geodiversity & Heritage
Sandford, Devon, UK

Diego Quiroga
Universidad San Francisco de Quito
Cumbaya, Ecuador

Raul Salazar
Biological Expeditions Galapagos
Puerto Baquerizo Moreno, Ecuador

ISSN 2363-765X ISSN 2363-7668 (electronic)
Geoheritage, Geoparks and Geotourism
ISBN 978-3-030-05914-9 ISBN 978-3-030-05915-6 (eBook)
https://doi.org/10.1007/978-3-030-05915-6

Library of Congress Control Number: 2018964920

This Springer imprint is published by the registered company Springer Nature Switzerland AG
The registered company address is: Gewerbestrasse 11, 6330 Cham, Switzerland

Contents

Introduction—Setting the Scene

1

When Charles Darwin stepped out of one of HMS Beagle's longboats onto San Cristobal Island on September 15th 1835—a Tuesday morning—his immediate impression was that "*nothing could be less inviting*": "*A broken field of black basaltic lava, thrown into the most rugged waves, and crossed by great fissures, is everywhere covered by stunted, sun-burnt brushwood, which shows little signs of life*" (Fig. 1.1).

Very soon afterwards, however, he began to realise that amongst the "*wretched looking little weeds*' and the "*antediluvian*" animals, "*the natural history of these islands is eminently curious, and well deserves attention*." The rest, as they say, is history—although at times maybe even mythology…

But it was also as a geologist that Darwin had joined the crew of the Beagle near four years earlier—although in the earlier 19th century, the distinction between the various natural sciences, such as zoology, botany and geology, was often blurred within a general concept of the 'natural historian'. He was strongly influenced by the Charles Lyell's, *Principles of Geology*, first published in 1830 a copy of which was presented to him by the Beagle's captain, Robert Fitzroy at the beginning of the voyage. In a geological context, Lyells's book is often considered to be as monumental in the development of science as Darwin's iconic *Origin of Species* of 1859. Through brilliant observation, Lyell had broken away from a literal, biblical interpretation of Creation, and had concluded that by observing natural processes *at the present time*, one could interpret the features one sees preserved in the geological record, literally "*the present is the key to the past*" (Fig. 1.2).

Lyell's contention that the rocks and landscapes we see today are a result of ongoing geological processes, leading to gradual changes over vast spans of time, were clearly fundamental to Darwin's explanation of the origins of the variety of life that he observed during his voyage on the Beagle (of which the Galapagos were only *part* of the puzzle…). However, Lyell did refer to the Galapagos in several places, including as examples of islands of volcanic origin—and this no doubt would have inspired the young Darwin. As well as his much more famous collections and observations of the wildlife of the islands, Darwin also collected a suite of rocks samples—now housed in the collections of the Sedgwick Museum of the University of Cambridge, England—which informed his later reports and theories. But he was not the only rapporteur on the geology of the islands, and many references to Galapagean volcanoes are scattered through 19th century and earlier 20th century scientific literature. But as with Darwin's geological notes, much of this is submerged, almost without trace, under a vast ecological literature, characteristically picking up where Darwin left of in his observations about the origins and evolution of the island's unique flora and fauna (Fig. 1.3).

Nevertheless, without an understanding of the origins and *geological* evolution of the islands, most of these evolutionary studies lack context—indeed, as the two themes are so intimately related, any study that fails to adequately consider this context may be fundamentally flawed. However, this geological literature inhabits an almost entirely independent publication world, with suites of scientific journals dedicated to volcanology and related themes offering little opportunity for cross-over, whilst ecological and evolutionary studies, with their own journals and populist outlets, always maintain the highest profile. In this volume, however, it is our intention to redress this imbalance, and provide a modern interpretation of the origin and evolution of the islands—informed by the latest geoscience research—in a form which will remain accessible, not only to ecologists, but also to the wide range of other visitors that the islands now receive, attracted by their its iconic status.

However, with increasing numbers of visitors, there is a need to provide a greater range of activities and opportunities for visits and the spectacular geological heritage of the Galapagos Islands can provide such a resource. And crucially, as the majority of geological features are extremely robust, there is a potential to use them to reduce some of the pressure on some ecologically more sensitive sites, as part of

D. Kelley et al., *In the Footsteps of Darwin: Geoheritage, Geotourism and Conservation in the Galapagos Islands*, Geoheritage, Geoparks and Geotourism, https://doi.org/10.1007/978-3-030-05915-6_1

Fig. 1.1 Darwin's map of the Galapagos islands, from his *Journal of Researches into the Geology and Natural History of the Various Countries Visited by H.M.S. Beagle*, 1835 and later editions, San Cristobal is Chatham Island

Fig. 1.2 HMS Beagle on the Strait of Magellan, South America (from the '*Journal*')

Fig. 1.3 Charles Darwin in the late 1830s—a changed man, soon after his return to England after the voyage of the Beagle (from https://commons.wikimedia.org/wiki/File:Charles_Darwin_by_G._Richmond.png)

the development of a more holistic approach to 'ecotourism' across the islands. Indeed, the whole issue of sustainability and tourism in the Galapagos has become a major theme in the published literature on the islands—now even appearing to swamp the number of evolutionary and ecological studies. A major theme within this literature is inevitably the delicate balance between tourism and conservation, with some proponents supporting increased controls and restrictions, whilst others promote a much more inclusive model that involves the island's resident population.

And this is where we see geological heritage as potentially being fundamental to future developments. Taking the highly successive model of UNESCO Global Geoparks as a framework for truly sustainable development (and with genuine benefits for the resident population), we argue that by developing the Galapagos' potential as both a '*geo-touristic*' and '*geoeducational*' destination, one can help consolidate a different style of tourism, in which education and understanding become as important as experience and entertainment. In this context, geological sites could become as important to as ecological sites—with visitors as keen to learn about the unique and spectacularly visible geology as they are to learn about the unique ecology and its origins. To such ends, we also provide an account of all the geological features visitors to the islands are likely to experience—or alternatively, might even wish to track down!

2 The Geology and Geodiversity of the Galapagos Islands

2.1 The Concept of Geodiversity and Geoheritage

2.1.1 Introduction

A desire to safeguard and protect aspects of our culture and environment for the benefit of future generations is probably as ancient as any other aspect of human consciousness. In the simplest form, this is represented by the process of handing down material possessions, both portable objects and defined geographical areas, to offspring, to help secure their future survival, both in terms of natural resources and the tools to exploit them—but also as tokens or symbols of social status and control, through which others might be invoked to carry out such tasks for the bearer's benefit. This concept of inheritance, or 'heritage' is a fundamental aspect of all human cultures.

Intertwined with this process is a notion, whether formally acknowledged or not, of what would now be termed 'ecosystem services', or the contribution of natural processes and systems to the well-being of human societies. Such benefits would include maintaining supplies of fresh drinking water, populations of wild game and other resources, such as wood, for construction. In most early societies, the enforcement of hunting restrictions, tree-cutting and water supply would have been primarily for the benefit of a ruling elite, but any management of a natural environment to maintain a *sustainable* access to a resource, must be considered as a form of 'conservation'. As with any general concept of inheritance, 'private' hunting reserves will have existed since the beginnings of settled, hierarchical human societies with the aim of maintaining—or sustaining—viable populations of, literally, target species—but also benefitting many other species. From the Middle Ages, at least, in Europe, formal 'Hunting Parks' were established by aristocracy and monarchy and ruthlessly guarded. Remarkably, the management of some has continued in some form to the present day, 'preserving' unique vestiges of ancient European landscapes and wildlife. Notable examples include the New Forest in southern England—now a National [landscape] Park (http://www.newforestnpa.gov.uk/info), with a Medieval patchwork-style landscape with areas of ancient oak forest (including *Quercus robur* L.), heathland and pasture, developed over Paleogene sands and mudrocks—but first established as a royal hunting forest in 1079, and the Białowieża National Park, in eastern Poland (https://bpn.com.pl/), a former hunting preserve first established over 800 years ago, which preserves one of the largest surviving remnants of indigenous European forest with a stronghold population of European Bison (*Bison bonasus* L.).

The development of a more 'altruistic' attitude to maintaining a *natural* environment certainly parallels an awareness of the destructive potential of human activities as large scale industrialization became established from the late 18th Century. Nevertheless, early campaigns to save natural features, such as landscapes, are likely to have had more to do with the aesthetic sensitivities of an intellectual elite of the contemporary society, than any actual concern for the natural environment, even indigenous cultures. Early examples included the protection of the forests of Fonatainebleu near Paris, France, from 1861, largely due to its artistic fame (de Wever et al. 2015), the saving from quarrying of the Cheesewring granite tor in Cornwall, south-western Britain in the 1870s (Dunkin 1870) as a culturally significant landmark and the halting of quarrying of Edinburgh's (Scotland) Salisbury Crag in 1845—a famous landscape feature overlooking the city (Thomas and Warren 2008) (Fig. 2.1).

Appreciation of the need to protect 'nature' from continuous human intervention and damage gradually developed in the latter part of the 19th century, as epitomized by the development of the US national parks movement. This campaign achieved its first success in the legal protection of the Yosemite Valley in California in 1864, followed by the world's first designated *National* Park at Yellowstone in 1872 (Gray 2004, pp. 176–179). Although, nature conservation, including of the famous geological features and systems such as the 'Old Faithful' geyser, was a core part of the aims for protecting the latter area, working with the

D. Kelley et al., *In the Footsteps of Darwin: Geoheritage, Geotourism and Conservation in the Galapagos Islands*, Geoheritage, Geoparks and Geotourism, https://doi.org/10.1007/978-3-030-05915-6_2

Fig. 2.1 The Cheesewring, a natural granite tor saved from quarrying by a local campaign supported in the scientific literature (Dunkin 1870); Bodmin Moor, Cornwall. *Photo* K. N. Page

area's original inhabitants (and 'owners') was not. As a result, around one million hectares were *"…reserved and withdrawn from settlement, occupancy* [etc.]" (Gray, *loc. cit.*) starting a trend in which 'protection' has all to commonly been synonymous with the removal of indigenous, or at least established, human populations.

However, with the development of systematic biological sciences and the realization of the fragility and vulnerability of natural systems—as indeed Darwin himself was probably one of the first to grasp—a more objective and scientific justification for protecting the diversity of natural features and systems could be articulated.

Initially, much of this effort has concentrated on the vulnerable living features, such as iconic species (e.g. the Giant Panda, *Ailuropoda melanoleuca*) and ecosystems (wetlands, forest, etc.), but in the later part of the 20th century, systematic programs for the selection and legal protection of geological and geomorphological features have also developed, especially across Europe, in part due the efforts of the NGO, ProGEO, The European Association for the Protection of the Geological Heritage (Wimbledon and Smith-Meyer 2012; www.progeo.se). Nevertheless, there are many earlier, independent examples of scientifically-motivated 'geoconservation' successes, such as the Upper Carboniferous 'Fossil Forest' in Victoria Park, Glasgow (Scotland) from 1887 (Cleal and Thomas 1995) and the declaration of the 'Petrified Forest National Monument' in Arizona, (USA) in 1906, with its Permian silicified tree trunks (Thomas and Warren 2008).

Ironically, some of the greatest early successes in developing a systematic approach to nature conservation—in its most general sense—have been in 'developed' countries, where much of the original biodiversity had already been damaged or destroyed by agricultural and industrial development. However, in regions still developing economically, massive loss of native species and habitats still takes place, often ironically to supply those areas which may have already destroyed many of their own natural features—but might have rigorous conservation systems in place to safeguard what still survives. Within this scenario, campaigns of course exist to protect the natural features of the regions now under intense exploitation. But some uncomfortable themes emerge—having damaged so much of their own environment for the material benefit of their societies, what right can so-called 'developed' countries have to dictate to others trying to achieve a similar, or at least 'better' standard of living for their own inhabitants? (Page 2018).

And so to the Galápagos, where such conflicts are very real, with a resident—although not perhaps 'indigenous'—population is confronted with the controls and restrictions promoted and enforced by 'outsiders', both from mainland Ecuador and internationally. Much has been written in recent years about such conflicts in the archipelago (Quiroga 2013; Powell and Ham 2008; Epler 2007, etc.), as discussed in Chap. 4. Indeed as a model of attempts at balancing such issues in a 'sustainable' way, it could be argued that the region is today perhaps as relevant as a global case study as it is for evolutionary processes. However, with a concentration on the most sensitive aspects of its natural environment, living species and ecosystems, some of the tremendous potential for sustainable economic development in the Galapagos has perhaps been missed.

This resource which has received scant attention is the islands' 'geodiversity', the natural, but non-living foundation of all ecosystems, but also very much part of a range of natural processes. It is the relative robustness, and ubiquitous character of this resource, combined with the dramatic features and stories it reveals, that give it so much potential for further sustainable development as a resource for visitors and for the benefit of resident populations.

2.1.2 What Is Geodiversity?

A key factor in promoting biological conservation globally has always been the establishment of international conventions, perhaps most significantly including the *Convention on Biological Diversity,* signed by 159 governments at the *Earth Summit*, which took place in Rio de Janeiro in 1992 (https://www.cbd.int/convention/). The Convention came into force on 29 December 1993 and was the first treaty to provide a framework for biological conservation internationally. It recognized, for the first time, that the conservation of biological diversity is "*a common concern of humankind*" and an integral part of the social and economic development process. Crucially, the agreement covered all ecosystems, species, and genetic resources. It also linked traditional conservation efforts to the economic goal of using biological resources sustainably and called for the creation and enforcement of national strategies and action plans to conserve, protect and enhance biological diversity (www.biodiv.org).

As a consequence of the adoption of the 'Rio Convention', the term 'biodiversity' has become well established in discussions of the conservation of species and habitats. It is, therefore, no accident that proponents of the need to conserve geological (e.g. rocks) and geomorphological (e.g. landforms) features began to use the term '*geodiversity*' from the mid 1990s—in a somewhat blatant attempt to exploit a renewed international and national interest in nature conservation. There appears to be no clear agreement as to who first used the term, however, but it is likely to have originated independently in several places (cf. Gray 2004, p. 6) as geologists realized the potential, political power of the concept of 'biodiversity'.

Crucially, however, fundamental links do exist between bio- and geodiversity, especially at the ecosystem level and the sustainable use of natural resources as promoted by the 'Rio Convention' really does include geological resources. So, in reality, it could be argued that the latter actually paved the way for a greater consideration of those natural features once considered as no more than '*inanimate nature*'. Things are changing, however, and due to considerable effort by geoscientists and conservation organizations, both nationally and internationally, the protection of geological and geomorphological heritage is now beginning to become established as a core part of nature conservation practice in many countries. This intimate relationship between aspects of ecology and geology is also beginning to be appreciated as concepts of 'eco-'and 'geosystem services' develop (Diaz et al. 2015; Gray 2013; van Ree and Beukering 2016; van Wyk de Vries et al. 2017).

Nevertheless, there still remains considerable confusion as to the exact meaning of the term *geodiversity*, and its relationship to conservation practice. The most detailed discussion on the concept remains Murray Gray's textbook '*Geodiversity: valuing and conserving abiotic nature*' (2004; second edition 2013), in which the author reviews previous definitions including those of Johansson (2000), Stanley (2001) and adopts the, much quoted, following:

"*The natural range (diversity) of geological (rocks, minerals, fossils), geomorphological (land form, processes) and soil features. It includes their assemblages, relationships, properties, interpretations and systems.*"

This definition is expanded in discussion within the original book to include geological materials in a cultural and economic context, for instance as building stones, ornaments and jewellery—following actual practice in the UK which predates Gray's definition—and is perhaps best revised as:

"*The term 'geodiversity' encompasses all aspects of the natural non-living materials and processes that formed our planet and continue to shape both its interior and surface today. This broad definition not only includes geological materials (such as modern sediments, rocks, minerals, meteorites and fossils), the processes that formed them (including by rivers and volcanic activity) and the landforms created by such processes (for instance cliffs and glacier-cut valleys), it also includes Earth materials removed from a natural to a cultural context, for instance to museums or used as building stones or in jewelry*".

[International Commission of Geoheritage: Terms of Reference 2018; http://geoheritage-iugs.mnhn.fr]

In addition, a plethora of other related terms have also appeared, and a copious amount of recent literature has been devoted to further and often increasingly and unnecessarily complex definitions—hence some more discussion is essential here.

Perhaps the most confused usage is of the term '*Geoheritage*' itself, with attempts to rigorously define it as some subsection of a broader geodiversity resource. In reality, however, the term can be equally appropriate in both a restricted and a general sense. In the latter context, it can be synonymous with geodiversity, as part of a broad natural, geological resource, for instance as discussed by Brocx and Semeniuk (2007) and Brocx (2008). Alternatively, it can also be used to describe a part of this broad resource, specifically selected for safeguard for future generations (e.g. in Sharples 2002—the criteria for this selection, however, will inevitably show different degrees of scientific objectivity, as discussed in Sect. 2.2.3 below). The context in which the term is used, however, should make its meaning apparent in different accounts.

When it comes to managing an identified, even 'selected', geological heritage for present and future generations various terms have been used. In the context of site-based '*nature conservation*' (a concept which should *always* encompass the management of both ecological and geological sites and features, including geomorphological), the once widely used term '*Earth science conservation*' (e.g. in NCC 1990, etc.) is misleading as it implies the conservation of geoscience activity rather than sites of *geoscientific* interest as studied by *geoscientists*. '*Geological conservation*' or '*Geoconservation*' are more appropriate terms, however, but '*Earth heritage conservation*' (e.g. in Wilson 1994) is perhaps the most explicit, but now rarely used. All three terms are interchangeable, however, and should be used to encompass the full range of geological *and* geomorphological features and processes which naturally occur on planet Earth and which can be considered 'worthy' of protection for future generations.

The term '*Earth resource conservation*', however, should refer only to the wise use of economically and culturally important geological resources (e.g. in Wilson 1994, pp. 156–7).

Another much abused term is that of '*Geotourism*', with definitions often trying to restrict its use to only those activities connected with a geological feature which might be educationally inspired (as discussed in Chap. 4. Nevertheless, any visitor to an area specifically to use or just view any geological features must also be considered to be 'tourist'—even geoscientists and rock climbers—especially if they purchase supplies and services from local suppliers. It is in this context that the term '*geotourism*' is used here with regard to Galapagos, as virtually all visitors inspired by geological features—whether it be a primary motivation or incidental to some other—will be visitors to the islands and hence will purchase supplies and services locally, and hence will contribute to the local economy.

Various other '*geo-*' prefixed terms have also been used, or are likely to be used, but not always in a geological sense, with some such as '*georeferencing*' and '*geolocation*' being more geographical rather than related to some aspects of the Earth sciences. As with the terms discussed above, understanding their general meaning is useful, but attempts at a scientifically rigorous definition can be confusing and often counterproductive.

2.2 The Geological Context and Geodiversity of the Galapagos Archipelago

2.2.1 Plate Tectonic Processes and Features

In order to discuss the geoheritage of the Galapagos Islands, including all of the geologic and geomorphologic features that define them, it is first important to understand why they are there. This archipelago that is positioned 1,000 km off of the coast of South America exists due to large scale Earth processes at play in this part of the world, and furthermore, the shape, age and character of the islands are a result of the geologic history and ongoing geologic processes. To fully understand the modern geological structure of the Galapagos Islands, it is necessary to understand the origins of the archipelago as a result of plate tectonic processes—hence a brief review of this fundamental Earth-system process is necessary:

The Structure of the Earth and Mantle Convection

The Earth's outermost layer, the lithosphere is divided into 12 major and a number of minor separate plates (Fig. 2.2). The lithosphere comprises the Earth's crust and the uppermost part of the mantle which move together as rigid plates due to the slow flow of the 'warm' [$\sim$1,300 °C] and hence relatively ductile mantle rock below (Fig. 2.3).

The mantle flows as a soft *solid* through deformation as heat is transferred from the Earth's interior to the exterior through convection. As the rock at the base of the mantle is hotter and less dense than the cooler denser rock above, it becomes relatively buoyant causing it to rise towards the surface. As this warm, upwelling mantle rock rises upwards, it cools and hence becomes denser, eventually beginning to flow back down. This upward and downward flow of mantle rock creates convection cells in the mantle (Fig. 2.4), resulting in a 'current' within the upper mantle which effectively 'drags' the rigid lithospheric plates above across the Earth's surface.

The elucidation of this process of 'plate tectonics', which was introduced in the early 1960s (Deitz 1961; Vine and

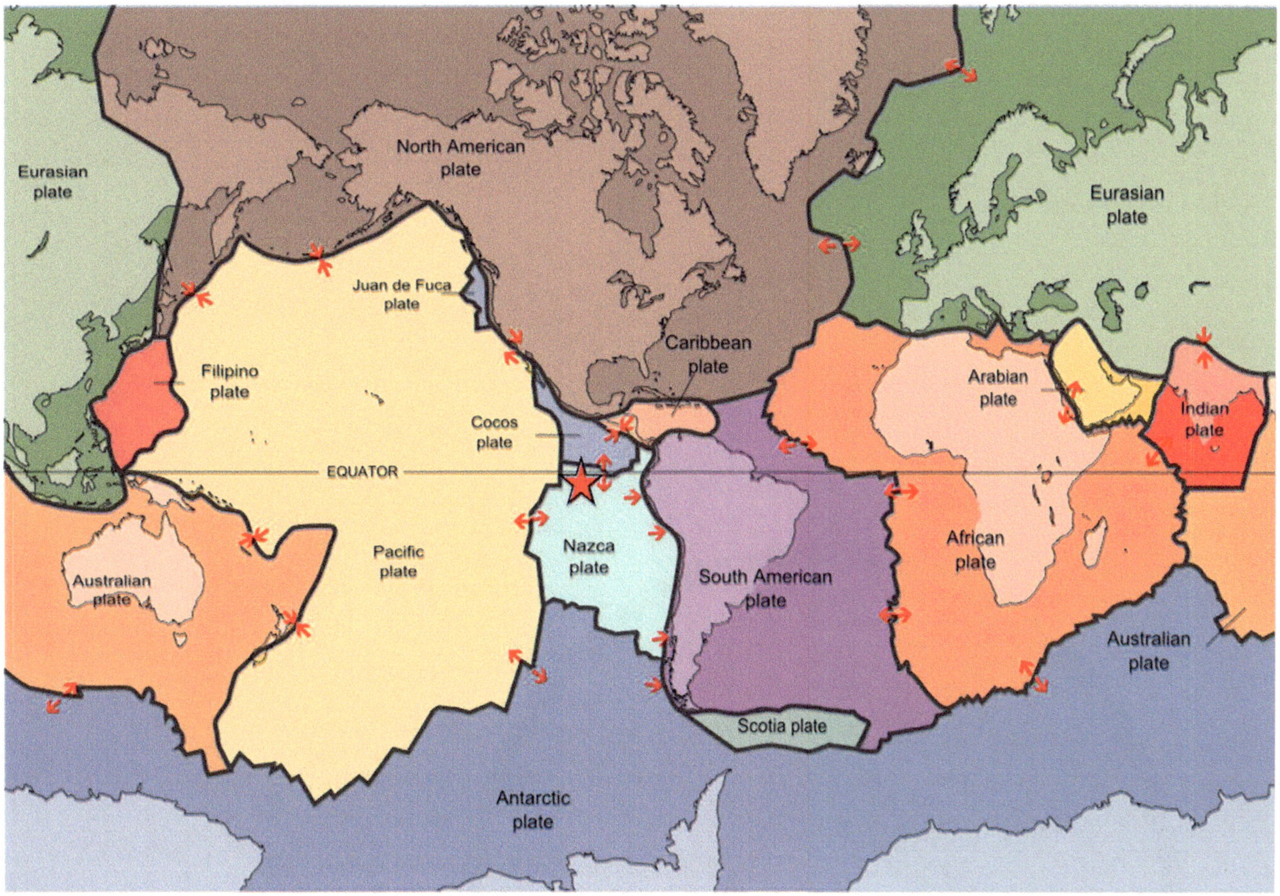

Fig. 2.2 Map of the Earth's tectonic plates. Black lines are boundaries between plates. Red arrows show relative direction of movement of plates at boundaries. Red star shows location of Galapagos Islands (modified from USGS map provided to the public domain)

Matthews 1963; Hess 1965), created one of the greatest revolutions in our understanding of the Earth as a system through the 1970s. It developed as diverse strands of evidence such as those related to continental drift, magnetic banding of oceanic crust, earthquake and volcano distribution and the palaeobiogeographic distribution of fossil faunas and floras were synthesized into a single unifying model (e.g. Wilson 1965, Morgan 1968, Le Pichon 1968). This model is fundamental to understanding the origins and evolution of the Galapagos Islands.

Divergent plate boundaries; Mid-Ocean Ridges

A boundary where two plates are moving in opposite directions away from their shared boundary is considered to be a constructive plate margin as it represents a zone where new oceanic crust forms. Such margins form where mantle convection cells act to bring warm mantle rock upward to toward the surface, and then act to pull the lithosphereic plates away from each other as flow of mantle rock separates (Figs. 2.3 and 2.4). As the warm mantle rock rises to near the surface, lower pressure is encountered, in turn lowering the melting point of the solid rock and initiating melting (e.g. Asimov et al. 2001).

The partial melting of the mantle in this zone (Fig. 2.5) generates magma which commences to rise toward the surface. This upward flow is not only driven by the lower density of the magma relative to the remaining unmelted mantle, but is also facilitated by strong tensional stresses acting on the plates of lithosphere as they are pulled in opposing directions, leading to development of cracks in the relatively brittle crust, through which the liquid can more readily ascend. Where the magma reaches the surface it erupts as lava and solidifies through cooling on the ocean floor. However, the rise of the magma is more often stalled within the cracks in the crust, where it cools and solidifies to create new crustal rock. As with erupted materials from the same source, these rocks have a basaltic mineralogical composition (see Sect. 2.2.5.4. below), which is the characteristic product of this plate tectonic setting.

This new oceanic crust is uniformly thin at around 8 km (White et al. 1995), much thinner than the continental crust (40–70 km) (Christensen and Mooney 1995) and forms the floors of the Earth's oceans. As these divergent boundaries,

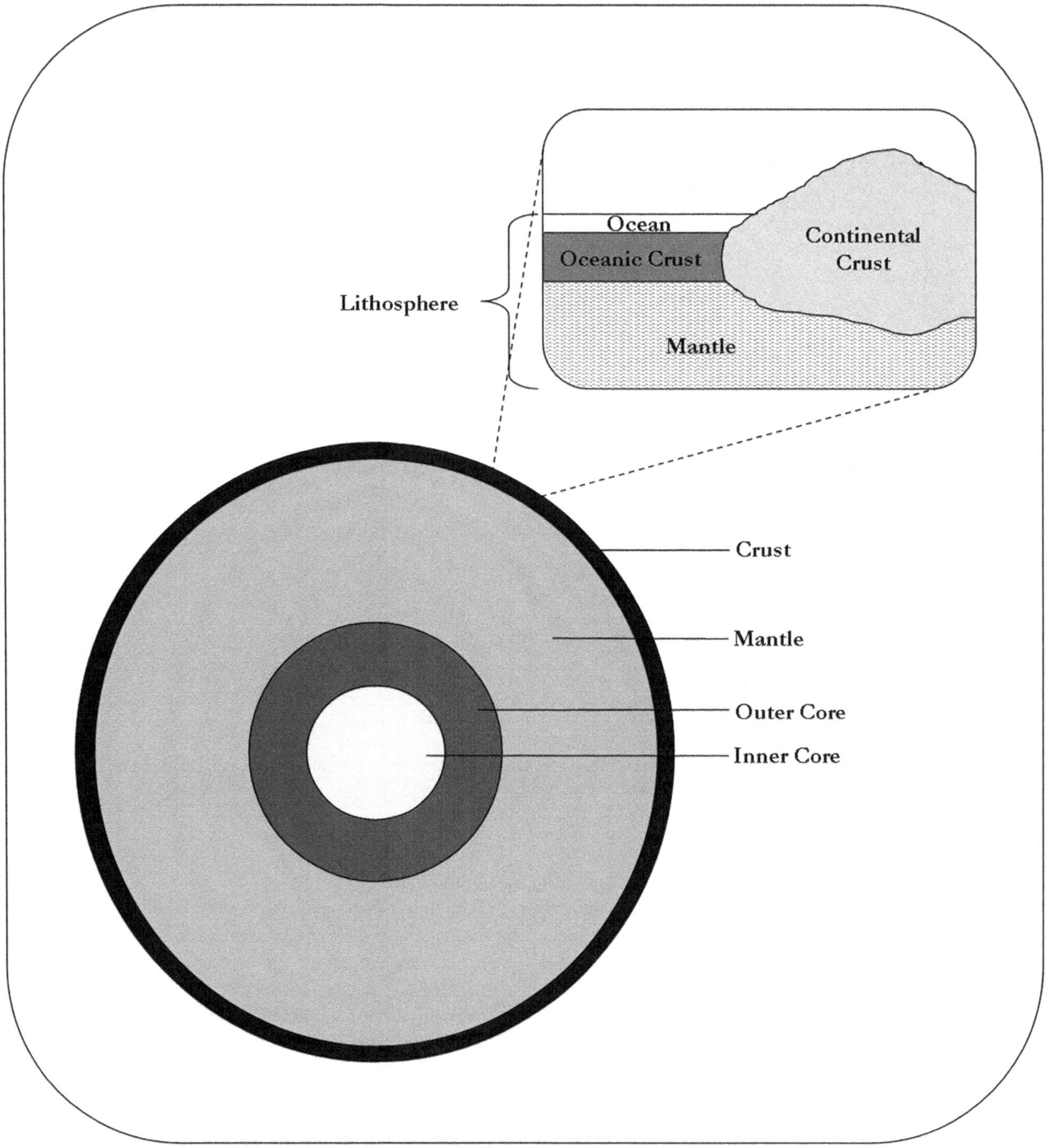

Fig. 2.3 The structure of the Earth's interior and outermost layers. The lithosphere comprises the crust and the rigid, uppermost portion of the mantle

which generate new oceanic crust are on the ocean floors, they are referred to as mid-ocean ridges (or 'MOR'). This slow, steady process leads to the ongoing creation of new crust, adding to each plate as they spread away from the MOR. The rate of this spreading varies between the Earth's MORs ranging from ~10 mm/year at the slow-spreading Mid-Atlantic Ridge to ~100 mm/year or more at the fast-spreading East Pacific Rise (Fig. 2.6) (Müller et al. 2008).

Convergent Boundaries; Subduction Zones

Over millions of years, the lithospheric plate comprising the more rigid upper mantle and the overlying oceanic crust

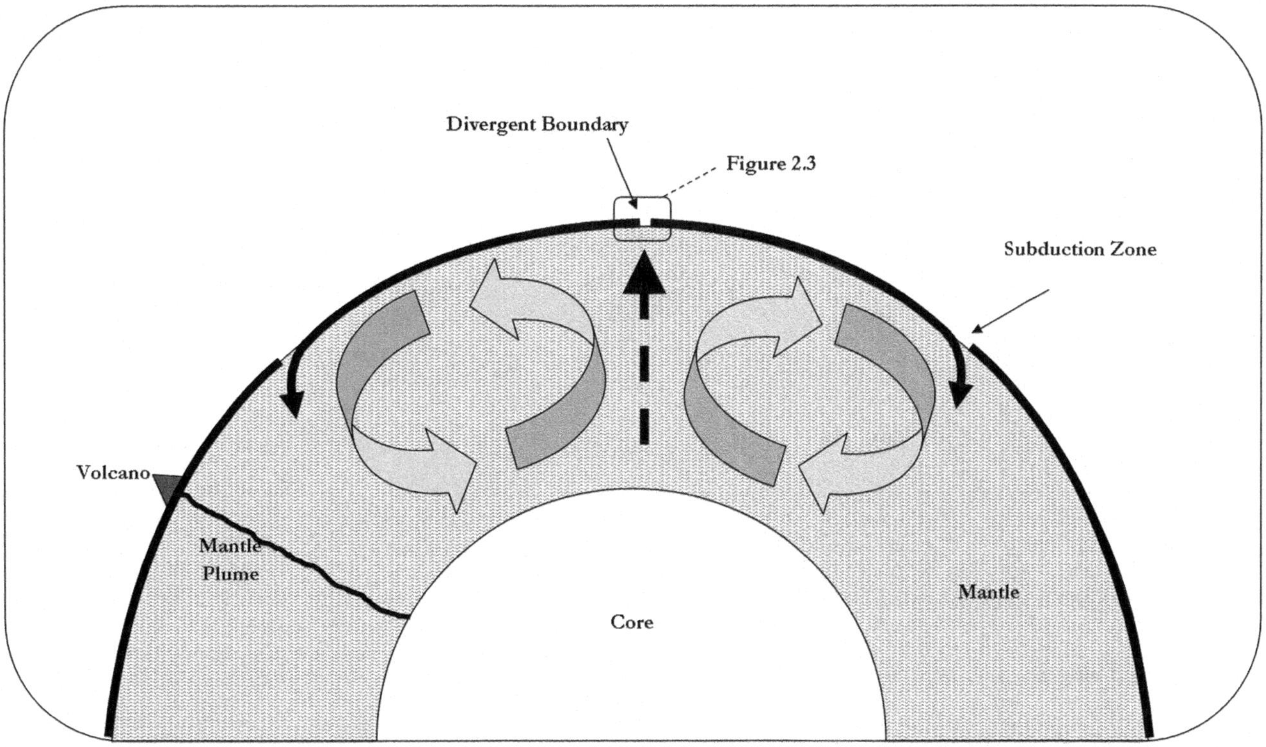

Fig. 2.4 Mantle convection: arrows show flow of mantle as it convects. The result is to move plates of lithosphere away from divergent boundaries where new crust is formed, towards subduction zones where it sinks back into the mantle. An independent mantle plume is also shown which functions largely separately from the current of convecting mantle rock

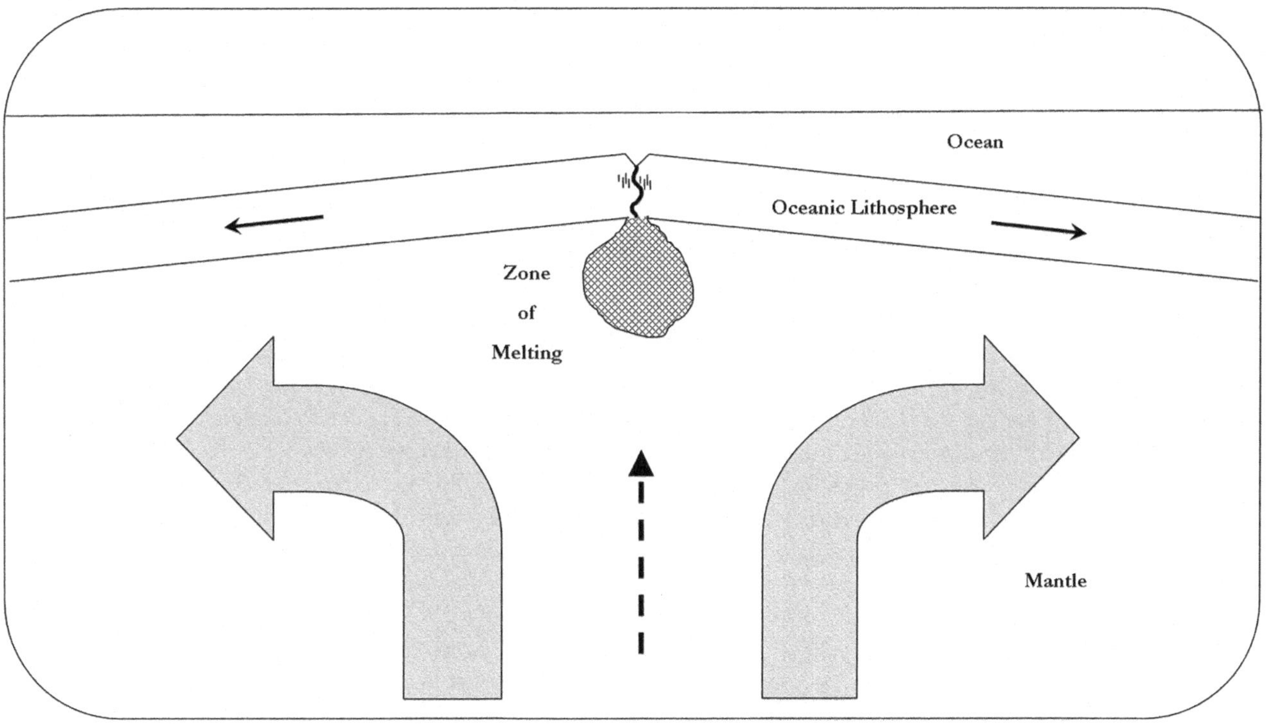

Fig. 2.5 The zone of partial melting below a Mid Ocean Ridge (MOR) that leads to the creation of new oceanic crust

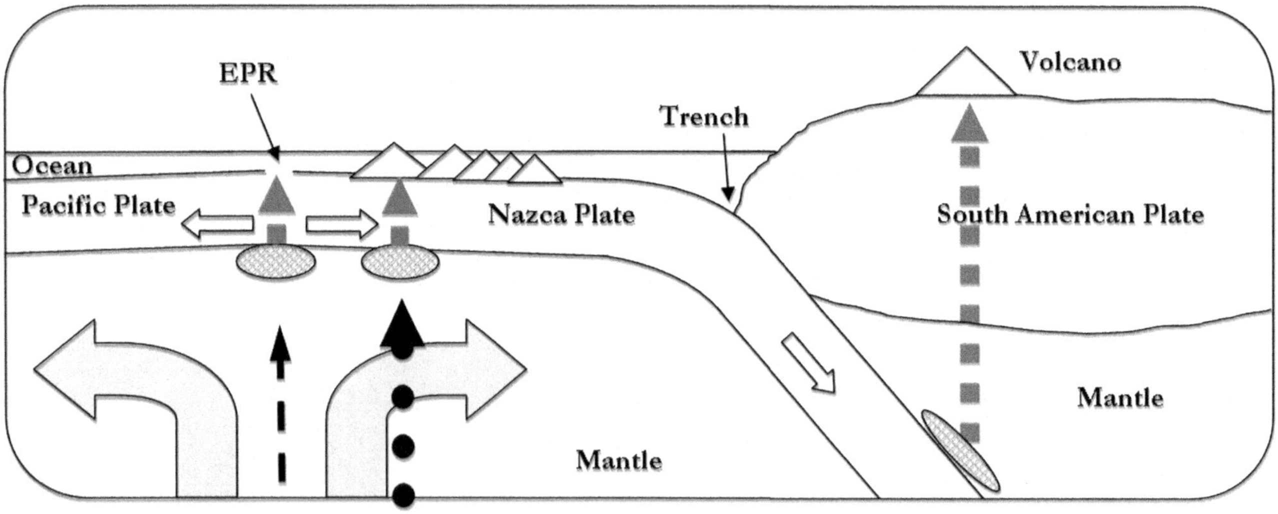

Fig. 2.6 Plate tectonic setting of the Galapagos Islands. The East Pacific Rise (EPR) is the MOR where the Pacific Plate and the Nazca Plate are created and spread away from each other (white arrows). The Nazca plate subducts beneath the South American Plate. Black dashed line show ascending warm mantle rock brought up through mantle convection. Large grey arrows show mantle convection path. The black arrow with round dots shows upwelling mantle rock of the Galapagos Mantle Plume. Zones of melting occur at the MOR, above the plume, and at the subduction zone. Grey dashed arrows show paths of ascending magma rising toward the surface. The triangles on the Nazca Plate indicate the volcanic islands of Galapagos created by eruptive activity above the Galapagos mantle plume

continues to move away from the MOR across the ocean, sinking into the mantle under its own weight as it cools and hence becomes denser. Eventually, however, as the Earth is not expanding, all oceanic lithosphere is destroyed as it sinks back down into the mantle along the zone where it is in collision with another plate at a convergent boundary (Figs. 2.4 and 2.6). As a tectonic plate descends, or subducts, a zone of melting is created, due to dehydration of the oceanic crust under increasing heat and pressure. When water from this dehydration is introduced to the mantle rock above the subducted slab, the melting temperature of the rock is lowered, allowing for formation of magma. The resulting magma produced along these subduction zones is, however, distinct in composition from that associated with MORs. The magmas formed in the mantle above the subducted slab may have some input from melting of the basaltic oceanic crust and/or the oceanic sediments. This magma will also rise toward the surface due to relative buoyancy. During this long ascent through the mantle and crust, the magma will evolve from a more mafic toward a more felsic composition (Table 2.1) by partial crystallization, and by melting in some of the surrounding rock.

Some of the magma will ascend all the way to the surface where it will erupt from the chains of volcanoes that characteristically develop above subduction zones (Fig. 2.6). Unlike the typically relatively 'gentle' eruptions of the more fluid basaltic lavas associated with MORs, the higher silica (i.e. quartz) content of the magmas produced above subduction zones produces more viscous lavas, such as andesites (Table 2.1), with a much more explosive eruptive character.

Where the oceanic lithosphere of one plate descends beneath a neighboring tectonic plate, a deep 'trench' on the ocean floor is typically developed; such trenches can be up to 11,000 m deep (Fisher and Hess 1963). The rate of subduction differs in locations around the world, but is on the same order of spreading rate at MORs and velocity of oceanic plate movement, being 10–100 mm/year.

Transform Boundaries

Not all margins between tectonic plates are constructive or destructive. At others, two plates simply 'slide' past each other along fault zones. These 'conservative' plate margins, or Transform Boundaries, are a consequence of the geometry of the Earth's tectonic plates—and the Earth's spherical surface—and allow the process of crust formation (at Mid Ocean Ridges) and destruction (in Subduction Zones) to continue without buckling the planetary surface. Perhaps the most famous—or infamous—of all constructive plate margins is marked by the San Andreas fault zone, on the western sea-board of North America. Along this margin the Pacific Plate to the west is moving northwestwards at around 30–50 mm/year relatively to the North American plate to the east (DeMets et al. 1987; Titus et al. 2005) (Fig. 2.2).

Table 2.1 Relationships of selected chemical and physical characteristics of lava based on the three primary compositional categories

Magma type	Basaltic	Andesitic	Rhyolitic
Plate tectonic boundary	Divergent, mantle plumes	Subduction zones	
Eruption temperatures (°C)	1,200	1,000	800
Relative viscosity	Low (like catsup)	Intermediate	High (like toothpaste)
SiO_2 content (weight %)	45–55	55–65	65–75
Significant chemical characteristics	High in Fe, Mg, Ca; Low in Na, K		High in Na, K; Low in Mg, Fe, Ca
Relative dissolved volatile content	Low	Intermediate	High
Explosivity during eruption	Low	Intermediate	High
Type of volcano	Shield, cinder cone, tuff cone	Composite	Collapse Caldera
Dominant minerals that form upon cooling	Olivine, Pyroxene, Plagioclase	Pyroxene, amphibole, plagioclase	Quartz, Potassium Feldspar, Plagioclase Feldspar
Possible minor minerals formed	Amphibole	Biotite, potassium feldspar, quartz	Amphibole, Biotite, Muscovite
Name of igneous rock that forms upon cooling	Basalt	Andestite	Rhyolite

Mantle Plumes

One final important feature related to the plate tectonic setting of the Galapagos is a mantle plume. These large scale features form above areas of the lower mantle that are excessively hot. This heat causes the mantle rock in this zone to be less dense than the surrounding mantle and as a result a stream of upwelling warm rock flows toward the surface. Where this warm rock rises to an area of low pressure, near to the Earth's surface, a zone of partial melting develops, similar to the zone under the length of a MOR. This scenario creates another setting for volcanic activity, often known as a hotspot, although this term is often used interchangeably with 'mantle plume'. As well as the Galapagos, other well-known locations for mantle plumes are beneath Hawaii and Iceland—although not all mantle plumes are the same, however, in terms of activity and volcanic products.

There is vigorous discussion in the research community about many aspects of mantle plumes (e.g. Foulger 2011), but nevertheless, these are relatively small regions of voluminous volcanic activity located above anomalously warm regions of the mantle. These mantle plumes can persist for tens of millions of years, seemingly unaffected by the currents of mantle convection around them, expressing themselves at the same point on the Earth's surface while the tectonic plates move above them (Condie 2001), (as illustrated as such in Fig. 2.6). As a result, the volcanoes that are built upon the surface of a tectonic plate are continually carried away from the hotspot while new volcanoes are built. This leads to a chain of volcanoes tracing a line where the plate has gone over the hotspot with the oldest being located furthest from the hot spot and the youngest, active volcanoes directly above it presently.

Magma Evolution and Lava Types at Plate Tectonic Settings

MORs, subduction zones and mantle plumes all generate magmas, as they are the three possible settings within the Earth where partial melting of the mantle occurs. Only transform margins do not (normally) produce magma. The magma produced in the mantle is typically mafic in composition meaning that it has a relatively high abundance of the elements Mg, Fe, and Ca and a relatively low amount of Si, K, and Na (see Table 2.1). However, while magma ascends through the mantle and crust toward the surface, it evolves chemically toward a more felsic end member with lower percentages of Mg, Fe, and Ca and higher abundances of Si, K, and Na. This chemical evolution is due to interaction with the crust and the phased crystallization of minerals due to the reaching of their freezing points with gradual cooling of the magma during ascent. Such minerals preferentially use some of the chemical components of the magma and leave the rest behind, leading to a chemical evolution of the magma. Because the distance from the zone of melting to the surface is short at MORs, there is much less of this chemical evolution than at subduction zones where the magma rises through a great thickness of mantle and crust (Fig. 2.6). Therefore, the lavas erupting at MORs and hotspots are typically basaltic in composition, while the lavas erupted at volcanoes above subductions zones are andesitic to rhyolitic in composition.

When a more evolved felsic magma erupts from the surface it is thicker, or more viscous, leading to the types of very explosive ash column eruptions that we see from the large volcanoes in the Andes Mountains (Fig. 2.7c). These subduction zone magmas also have a higher amount of dissolved gasses in solution, resulting in the formation of a high volume of gas bubbles upon eruption under the low pressure conditions of the Earth's surface, adding further to the explosivity of these eruptions.

When a hotter, more 'primitive' magma erupts—such as above a Mid Ocean Ridge—its relative fluidity allows it to rapidly flow, spray or spatter. This basaltic lava is also typically erupted at ocean island hotspots such as the Galapagos Islands (see example eruption Fig. 2.7d).

Volcanic Rocks

These different magmas, ranging in chemical composition (in particular relative amounts of the elements Fe, Mg, Ca, K, Na, and Si), will form different minerals upon cooling and crystallization. Therefore, when a basaltic magma solidifies through crystallization, it will typically form the minerals plagioclase feldspar, pyroxene, and olivine. When a rhyolitic magma crystallizes, there are different chemical components available with which to form the crystals of the solid rock and therefore a different assemblage of minerals will result—in this case quartz, plagioclase feldspar, potassium feldspar, and biotite. The rocks that form as a result of crystallization of different types of lavas will have different names based on

(a) (b) (c) (d)

Fig. 2.7 **a** Sierra Negra, Isabella, Galapagos—a typical 'shield volcano' formed from the eruption of fluid basaltic lavas, hence the low profile and gentle slopes, rising to only 1000 m. *Photo* D. F. Kelley; **b** The classic Andean volcano rising to 5911 m above sea-level—Cotopaxi (Ecuador)—the product of ~20,000 years of explosive eruption of ash and lava of andesitic composition. *Photo* D. F. Kelley; **c** Explosive volcanic eruption producing an ash column at Tungurahua Volcano, Ecuador. *Photo* D. F. Kelley; **d** Non-explosive eruption producing low viscosity basaltic lavas flowing on the flank of Sierra Negra volcano, Galapagos. *Photo* Christian Saa

their mineral make-up. Mafic lavas form basalt; intermediate lavas form andesite; felsic lavas form rhyolite (see Table 2.1). Since these different types of lavas are related to the distance of ascent at the different plate tectonic boundaries, there are typical rock types produced at each setting. In the Andes Mountains of Ecuador, the volcanoes produce andesite, dacite, and rhyolite. In the hotspot setting of the Galapagos Islands, the lava is basaltic in composition and the resulting rock is also called basalt.

Eruptive Style

When lava erupts from a volcanic vent it can do so either effusively or explosively. The former refers to liquid that flows out of the vent directly onto the surrounding landscape. The latter refers to any eruptive activity that involves throwing lava through the air as it exits the vent. As lava ascends from depth, it rises from higher to lower pressure areas over a relatively short amount of time. This change in pressure affects the solubility of gasses within the magma, and degassing occurs as the liquid reaches the surface. Gasses such as water vapor, carbon dioxide, sulfur gasses and others come out of the solution of the liquid and become free gasses creating many bubbles in the lava—and hence an explosive tendency as the gasses escape. Explosivity can range from lava "spattering" out of the vent only feet into the air and landing quite locally around the vent to lava being pulverized into very fine particles and ejected 10s of thousands of feet into the upper atmosphere—for instance the infamous eruption of the Icelandic volcano Eyjafjallajökull which grounded international flights in 2010.

The amount of explosivity is a result of the amount of gasses that are coming out of solution while the magma is ascending in combination with the viscosity of the lava. By its chemical nature, basaltic lava does not release a great deal of gasses during eruption as compared to andesitic or rhyolitic lava. Also, as discussed above, basaltic lava is less viscous, and so the gas bubbles that do form in the liquid can escape relatively easily. Conversely, when a thick rhyolitic lava is ascending, with the gas bubbles forming in the liquid cannot escape, resulting in the very explosive eruptions in which the lava is pulverized during violent ejection from the vent.

Any resulting volcanic material that is created through an explosive eruption is given the term tephra. Tephra is further characterized based on the size of the particles that have flown and landed. Ash sized particles are those that are less than 2 mm in diameter. Those particles falling within the range 2–64 mm are referred to as lapilli, and larger clasts are called either blocks or bombs.

2.2.2 The Plate Tectonic Setting of the Galapagos Islands

The Galapagos archipelago is located in a position on the Nazca Plate (Fig. 2.8) where MORs, subduction, and a mantle plume are all at play. Before characterizing the islands in detail, it is useful to place them in their broader plate tectonic context.

The Nazca Plate is one of the Earth's tectonic plates that is entirely composed of oceanic lithosphere. This plate is being created by spreading activity along the East Pacific Rise (EPR) and then moves in an eastward direction, guided by mantle convection, before it is eventually subducted beneath the western margin of the South American Plate (Figs. 2.6 and 2.8), which, here contains a great thickness of continental crust. The magma that is generated by subduction of the Nazca Plate leads to the volcanism that has created the Andes Mountains and fuels the ongoing eruptive activity there. This lava that is erupted is predominantly andesitic in composition as is expected from the discussion in the previous section. In fact, the Andes Mountains are the type locality for this type of rock, thus the name—Andesite.

The Galapagos Islands' position on the Nazca Plate is due to volcanic activity above the Galapagos mantle plume. Through seismic imaging, this plume is known to extend to a depth of as much as 1,000 km (Montelli et al. 2004), while chemical and thermal investigation by Harpp et al. (2014) suggest its root could be as low as the base of the mantle (i.e. 3000+ km; see Fig. 2.3).

The initiation of volcanic activity above the Galapagos hotspot occurred around 95–72 million years ago (Hernle et al. 2002). This initial hotspot activity created a large plateau on the seafloor which has since migrated through plate tectonics to the present Caribbean Sea (Thompson et al. 2004). Approximately 22 million years ago, a major restructuring of plates along the western edge of the Americas took place, as the large, former Farallon Plate broke into smaller plates, including the Cocos and Nazca plates (Figs. 2.2 and 2.8). Over the next several million years, the boundary between these latter two plates evolved, eventually settling as a divergent boundary with an oblique angle between the eastward moving Nazca and northeastward moving Cocos (see Fig. 2.8).

The location of the plume and hotspot has been more or less coincident with this spreading center for most of the time since. It is possible to determine which portions of the plates have a history of interaction with the plume because the overactive volcanic activity has created a thickened area of the oceanic crust, and with the movement of the plate over the hotspot a ridge has been created. This is a result of the

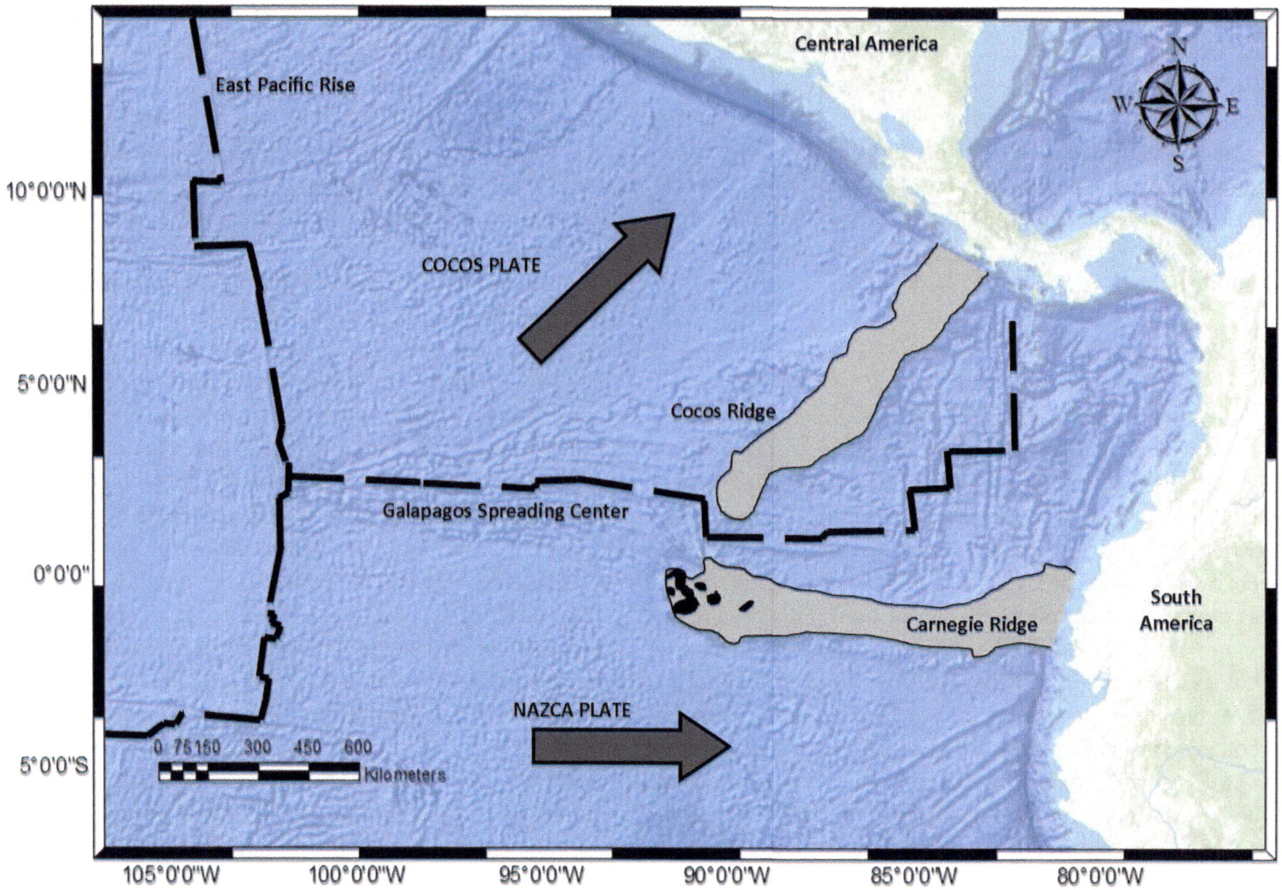

Fig. 2.8 Map showing the plate tectonic setting of the Galapagos Islands (in black). Long dashed line is the spreading boundary between Nazca, Cocos and Pacific plates. Subduction occurs at the boundary of continents where the trench can be seen in ocean floor shading of dark blue

magma generated at the base of the lithosphere above the plume ascending and being intruded or erupted onto the surface of the plate, hence thickening it. The Cocos Ridge has formed on the Cocos Plate and the Carnegie Ridge has formed on the Nazca Plate (Fig. 2.8). Portions of the Nazca Plate have been over the hotspot for the 22 million years since the initial plate rearrangement. The Carnegie Ridge contains some rock created over the past 20 million years (Meschede and Barkhausen 2001; Harpp and White 2001). The eastern most portion of the Carnegie Ridge, however, started to disappear through subduction beneath South America from around 2 million years ago (Gutscher et al. 1999; Meschede and Barckhausen 2001).

To the north of the Galapagos Islands, the Galapagos Ridge separates the Nazca Plate from the Cocos Plate (Fig. 2.8). While the two plates are not diverging directly away from each other, as is the case at a typical MOR, this is a spreading center none-the-less. From 20–7.5 million years ago, the Galapagos hotspot was centered on the Galapagos Ridge or under the Cocos Plate thus "leaving its mark" on the Cocos plate for that entire time span. Sustained volcanic activity at one place above the plume allows for the eventual building of volcanic edifices that are sufficiently high to be above sea level. When the edifice is not sufficiently high to rise above sea level, it is called a sea mount. The chain of sea mounts and islands on the Cocos Plate that were built by volcanic activity influenced by the Galapagos mantle plume collectively comprise the Cocos Ridge (Fig. 2.8). However, from around 7.5 million years ago until the present, the plume has centered under the Nazca plate (Barckhausen et al. 2001; Sallarès and Charvis 2003) resulting in the Carnegie Ridge.

The Galapagos platform is the western end of the Carnegie Ridge hotspot track. It has been built up through the most recent volcanic activity which persists today. The platform is a wide cluster of volcanic terraces creating an area of relatively low sea level (Geist et al. 2008a, b) that is elevated with respect to the average sea depth on the Nazca Plate (See Fig. 2.6). The islands are the highest portions of the platform, and thus are the only portions that are above sea level.

This portion of the Nazca Plate is also slightly elevated relative to the rest of the plate due to the buoyant upward force of the ascending mantle plume. Therefore, as the volcanic rocks of the Carnegie Ridge (including any islands) move eastward away from the hotspot, they subside relative to sea level. Additionally, as they age, they are eroded more and more. As a result, the islands over the hotspot eventually sink below sea level to become seamounts once they have migrated sufficiently far to the east (Geist 1996; Geist et al. 2014). Over the time span of the tens of millions of years that this process has continued, many islands will have existed and then disappeared to become the sea mounts of the Carnegie Ridge and the Cocos Ridge.

2.2.3 Volcanic Activity in the Galapagos Islands

Due to the eastward movement of the Nazca Plate, the most volcanically active islands with the youngest deposits of volcanic rock are at the western end of the archipelago. The Galapagos, however, are different in this hotspot track behavior than in a more typical situation, such as Hawaii, as there have been eruptions throughout the archipelago in relatively recent history, with the lava fields on the north-eastern half of San Cristobal island likely to be less than 1,000 years old (Geist et al. 1986). In addition, none of the volcanoes are significantly bigger or taller than the others and Isabela Island is only the largest due to the linking together of a series of 6 shield volcanoes. It seems that the head of the mantle plume could be 'stretched' eastward as it meets and is dragged by the base of the lithosphere of the Nazca Plate (White et al. 1993; Harpp and White 2001). This leads to most of the eruption of lava coming from the volcanoes of the western islands, but within a history of lava erupting all across the platform.

The volcanoes of the Galapagos Islands are quite active. According to the Smithsonian Institute's Global Volcanism Program (GVP), there have been over 65 confirmed eruptions from volcanoes in Galapagos since the year 1800. Most of these have occurred in the westernmost islands, which are currently located above the hotspot. A map of these volcanoes is provided in Fig. 2.9. These eruptions have included activity at La Cumbre Volcano on the island of Fernandina, as well as Cerro Azul, Sierra Negra, Alcedo, Wolf and Darwin Volcanoes on the Island of Isabela. There have been six eruptions between these two islands thus far in the 21st century.

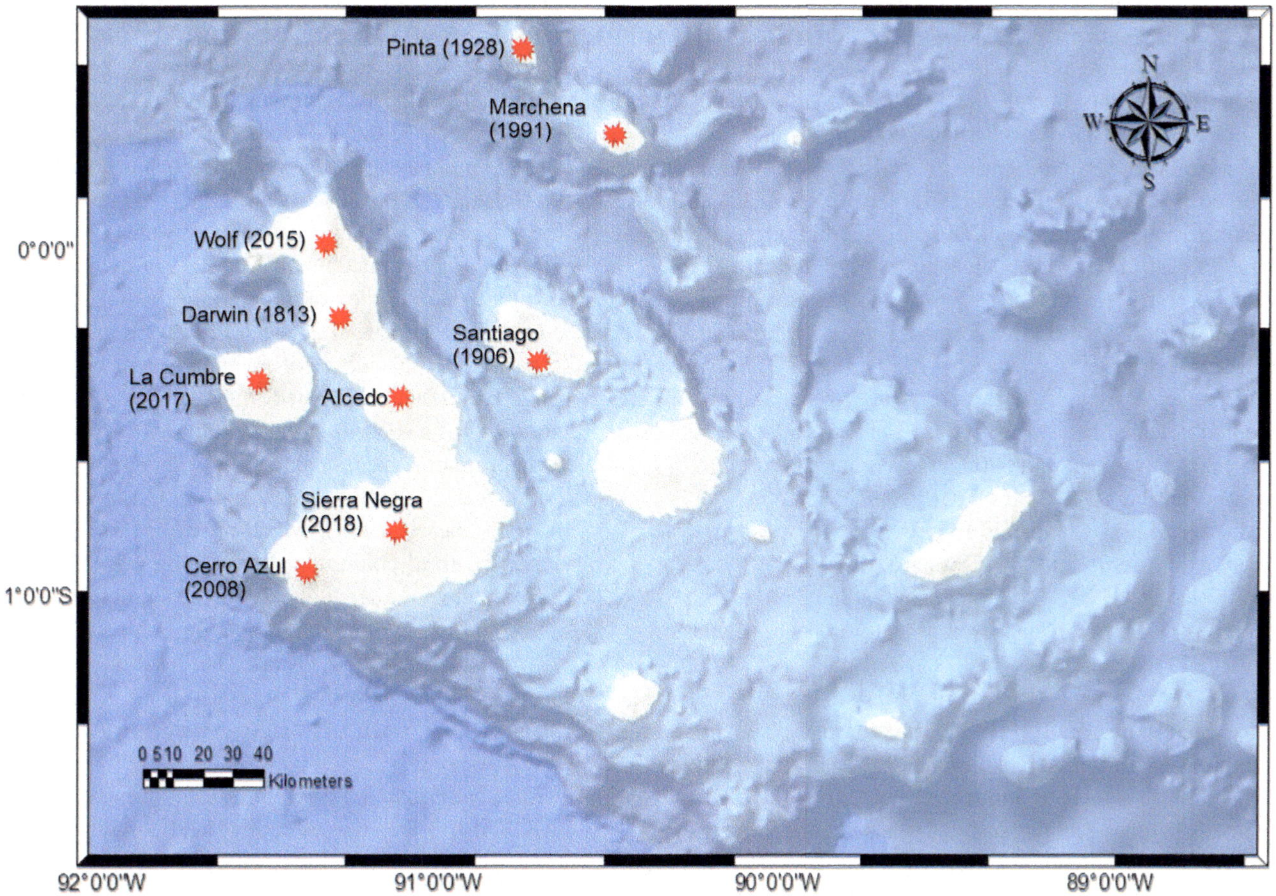

Fig. 2.9 Map of Galapagos Archipelago with red markers showing locations of active volcanoes. Along with the name of each volcano is the year of the most recent eruption

Fernandina is the most active of these volcanoes in recent decades, erupting in 1991, 1995, 2005, 2009, and 2017 (Source: GVP). However, as this volcano is on an uninhabited island, it poses no threat to resident human life or property. Likewise, most of the volcanoes on the large Isabela Island are located far from any settlement of people, but with the notable exception of Sierra Negra. The port town of Puerto Villamil is located on the distal flanks of Sierra Negra and the community of Santo Tomas in the highlands on the eastern flank. Considering the frequency of activity of Sierra Negra and the lava flows that can be seen to have flowed from the crater to the sea in the past (Reynolds et al. 1995), this volcano could potentially create a hazard for the roughly 2000 inhabitants of the island.

The second most recent eruption of Sierra Negra, in 2005, was well-observed and documented (Geist et al. 2008a, b). The eruption was sourced from the NE edge of the caldera, from which lava flowed out of the crater partially, but mostly into the crater covering much of the floor with a layer of basalt. This black color of the fresh basalt produced by this eruption sharply contrasts with the green vegetation on the crater walls, adding to the visual attraction of this popular tourist site (Fig. 2.10a). This ability to visualize the products of a recent eruption, along with the proximity to the Volcan Chico area, is aided by the possibility to drive from Puerto Villamil, including in tourist buses, close to the caldera hence making this one of the best sites in the islands to demonstrate an 'active geoheritage' to visitors.

Sierra Negra erupted again in 2018. This eruption was sourced from a series of fissures on the western and northwestern flanks of the volcano. The eruption was not observable from the accessible portions of the rim of the volcano, but lava could be easily observed from ships (Fig. 2.10b). This eruption posed no threat to residents or tourists

The Cerro Azul, Alcedo and Wolf volcanoes have all erupted at least once since 1990 on Isabela Island (Teasdale et al. 2005; Nauman et al. 2002; Geist et al. 2005), with the most recent being at Wolf in 2015 (IG-EPN 2015).

In the central Galapagos Islands, there have been relatively young eruptions on Santiago Island as well as the small islands of Pinta and Marchena. On Santiago, eruptions occurred in the 1600s and most recently in 1906 (Swanson et al. 1974). Pinta erupted in 1928 (McBirney and Williams 1969) and Marchena erupted in 1991 (GVP 1991). These three islands are part of the northern trending lineament that continues to Darwin and Wolf Islands. The eruptive activity along this lineament is influenced not only by the mantle plume that generated the hotspot, but also by the influence of the Galapagos Spreading Center (see Fig. 2.8) (Harpp and Geist 2002).

Moving to the eastern end of the Archipelago, and to San Cristobal Island, a strange and interesting volcanic history is encountered. This island, following the oceanic island hotspot model, is the oldest in terms of its construction, with rocks having been dated to around 2.3 Ma (Geist et al. 1986). These relatively old rocks are mainly found in the south-western half of the island which is a single, large shield volcano. The building of this volcano, and therefore this portion of the island, continued until around 660,000 years ago (Bailey 1976). The northeast half of the island is quite different, however, a contrast noted already by Darwin (1844), who described the many small craters and lava flows of this relatively flat and low-lying portion of the island. While there is currently no direct dating available for the age of these lavas, they are by appearance quite young. Geist et al. (1986) constrained the age of these eruptions to have been most likely between 1,000 and 150 years ago.

2.2.4 Relative Sea Level and the Galapagos Islands

Relative sea level is variable with time and so the location of the coastline of any given island is has varied significantly through time. This is due to four primary factors

(1) In the case of active volcanic islands such as the Galapagos, the islands continue to grow over time, albeit slowly, due to accumulation of new lava. This applies mainly to those islands that are at the western end of the chain and thus over the hotspot at a given point in time, but as discussed in the previous section, others islands can have activity as well. During an eruption, it is not uncommon for lava to flow primarily down one flank of the volcano and for that coast line to be expanded somewhat as the lava flows into the sea.
(2) Local uplift or subsidence of different islands or parts of islands can occur associated with the flux of magma into or out of a shallow storage chamber. Darwin noted elevated marine sedimentary deposits and shells at several meters above sea level on San Cristobal island in 1835 (Darwin 1844), which are likely to represent such a localized phase of uplift hoisting these deposits up from below the sea level. There are similar locations around the islands, including notably Bahia Urbina on the west side of Isabela Island. This location suddenly uplifted in 1954, extending the coast line ¾ mile out to sea, and lifting beach and shallow marine deposits up to 5 m above sea.
(3) As already discussed, as oceanic crust moves away from a MOR it subsides due to cooling and contraction and hence its increasing density (Hillier and Watts 2005). The same effect, although perhaps not as prominent occurs as the plate moves over and then away from a mantle plume (Detrick and Crough 1978; Clift 2005). As

Fig. 2.10 **a** Caldera floor and wall of Sierra Negra Volcano, Isabela Island. *Photo* D. F. Kelley; **b** 2018 Eruption of Sierra Negra. *Photo* Christian Saa

a result, the islands and the entire platform are continuously subsiding as they move eastward (Fig. 2.11b) (Ali and Aitchison 2014). Therefore, the portion of the Nazca plate that is moving away from the Galapagos hotspot and becoming part of the Carnegie ridge is riding at a lower and lower level, and hence any islands present will eventually slip below sea level. The highest sea mounts on the Carnegie Ridge are just to the east of the Galapagos Islands and are likely to have originally been islands before migrating away from the hotspot. The gradual subsidence and submergence of San Cristobal Island has been modeled and described by Ali and Aitchison (2014).

(4) **Eustatic sea level** change refers to change that is global. Sea levels change dramatically, as a result of natural climate cycles driven by the Earth's orbital cycles (e.g. 'Milankovitch cycles'), such as those driving the glacial—interglacial cycles of the last 2 million years or so (Miller et al. 2005). The timing of these eustatic changes is well constrained, with ice age periods creating low stands of the sea as water is locked up as polar ice, and warmer interglacial intervals melting polar caps leading to sea level rise. These low stands of sea level have occurred around every 100,000 years over the past 1 million years with the most recent peaking around 19,000 years ago, see so-called 'Late Glacial maximum' (see Fig. 2.11b).

All of these effects combine to determine the *relative* sea level, or where the coastline lies on the islands, at any point in time. Ali and Atchison (2014) modelled the consequences of effects 3 and 4 above over the Galapagos platform. In Fig. 2.11 from their work, the eustatic sea level fluctuations are combined with the subsidence effect that occurs as the plate moves away from the hotspot. It can be seen in Fig. 2.11b that older portions of the platform that are now as much as 210 m below sea level could have been at sea level when eustatic change, and the amount of subsidence that they have experienced are both considered.

As a point of reference, Fig. 2.11c shows what the Galapagos archipelago would look like today with a drop in

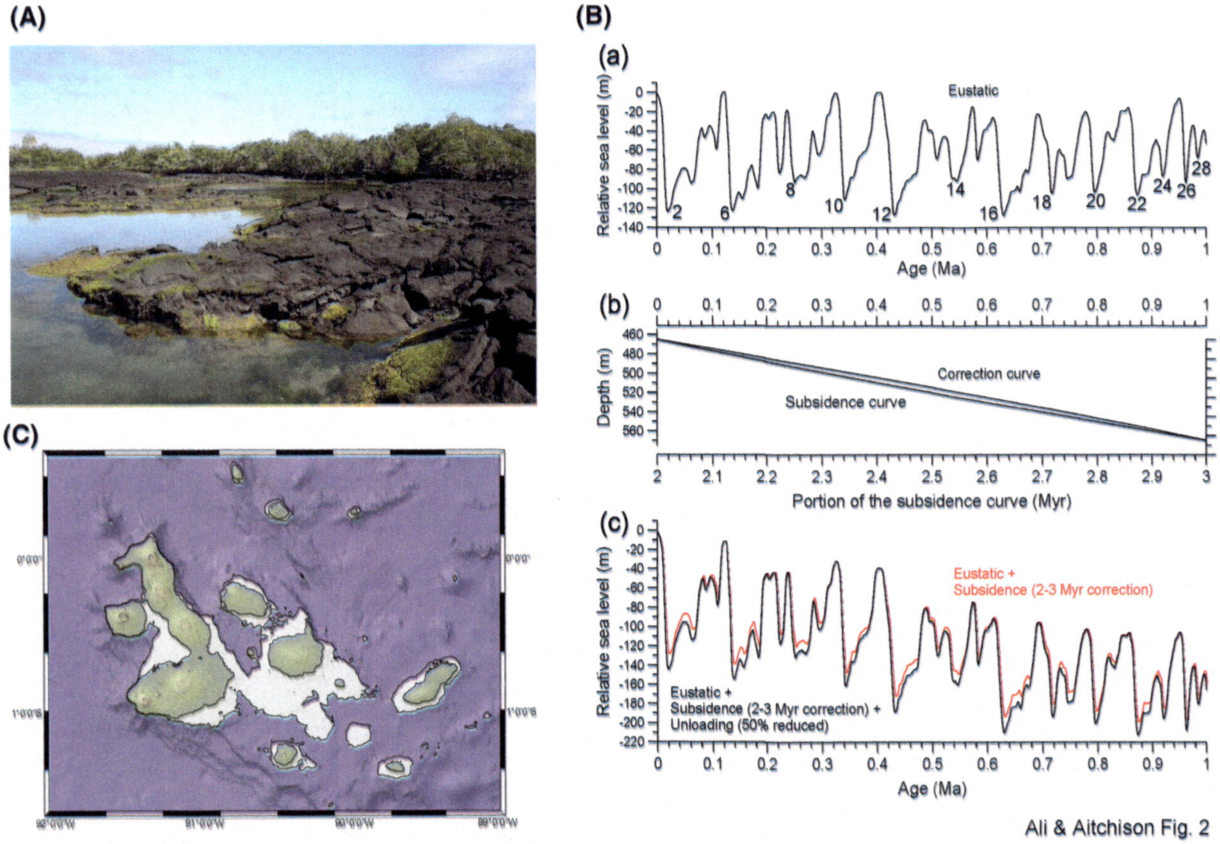

Fig. 2.11 **A** Punta Espinosa, Fernandina—shore platform of basaltic lava flow formed by an eruption in 1825 and raised above sea-level by a volcanic event in the 1970s. The former landing stage for tourist boats (inset) can just be seen by the mangroves at the back of the far pool. *Photo* K. N. Page; **B** From Ali and Aitchison 2014 *Journal of Biogeography*. (**a**) Shows eustatic sea level over the past 1 million years relative to present with even numbers indicating marine isotope stages; (**b**) showing calculated curves for sea floor subsidence with age; and (**c**) providing a combination of these two effects to give relative sea level for Galapagos during the last 1 million years; **C** map of the Galapagos Islands with today's shorelines in black outlining the green islands as well as the shoreline as it was with the sea level 210 m lower than today

sea level of 210 m. While the islands are continually changing in size and position, this demonstrates that it is quite likely that during times of low sea level a much greater portion of the platform was exposed, creating linkages between what are now separate islands. These low stands in sea level when the islands were coalesced lasted 5,000–10,000 years.

The appearance of new islands over the hotspot and disappearance of old islands through subsidence to the east as well as the connection and separation of islands due to sea level fluctuations have had important impacts on the mixing and isolation of animal species through time as is discussed by Ali and Aitchison (2014), and in Chap. 3 here.

2.2.5 The Volcanic Geodiversity of the Galapagos

The volcanic geology that has resulted from the plate tectonic setting of the Galapagos Islands provides the framework upon which the rest of the natural environment has been developed. While many residents and visitors to the islands hold the biological ecology at the forefront of their attention, the islands are first and foremost a volcanic island chain. Understanding the nature and origins of the geological features typical of this setting—in other words the island's *geodiversity*—as well as the underlying processes that have created them and still operate today, is crucial to understanding the past, present and future of these islands and their unique ecology. Above all, the Galapagos Islands are a classic example of basaltic ocean island volcanism above a mantle plume.

At a most basic level, when lava finds a vent from which to repeatedly erupt, more and more lava and tephra builds up around the vent, resulting in the growth of a volcano. In places where lava erupts from a vent only once, a much smaller feature is constructed, however. The geodiversity of the Galapagos Islands contains features spanning this spectrum from single eruption to many events over a long time span. Cumulatively, ongoing eruptions from many locations over ~ 5 million years have constructed the entire Galapagos platform. The islands, being the highest points on the platform are the volcanoes that have built up through many eruptions from each vent. The many small islets of the archipelago are either single eruption features, the remains of eroded structures formed by multiphase eruptions or a combination of these two. While there are many different types of eruptive features and structures, they are all the result of much larger scale, plate tectonic processes which have led to partial melting of the mantle to generate magma, which in turn ascended through the crust and ultimately erupted as lava on to the Earth's surface. Here we will consider the largest features first:

2.2.5.1 The Galapagos Platform

The platform itself is an important aspect of the geography of Galapagos. It sets this area apart from the surrounding region of the equatorial Pacific. The production of volcanic rock above the Galapagos mantle plume, has built up the current platform with quite steep sides in some places to an elevation that up to 500 m higher than the surrounding seafloor on the Nazca Plate (Geist et al. 2008a, b). This shallow water environment in the midst of the deep sea provides area for a rich marine biodiversity (as reviewed by Constant 2003). This diversity is the reason for the establishment and protection of the Galapagos Marine Reserve, and motivates many visitors who seek the snorkeling and scuba diving sites.

In addition, the platform has provided the land bridges between the islands during periods of relatively low sea level, which has had such an important impact on the evolution of animal and plant species across the islands.

2.2.5.2 Major Volcanic Landforms and Related Features

(a) **Shield volcanoes**—Each of the larger islands consists of one to several shield volcanoes. A shield volcano is a volcano that has a large diameter relative to its height. From a distance, these features can resemble a warrior's battle shield. The shape is a direct consequence of the low viscosity of basaltic lava, which when erupted subaerially, can cover long distances before cooling and solidifying. However, over many eruptions, successive lava flows can build up a structure to significant elevations, but gradually.

The maximum elevation of the shield volcano that makes the southern half of the island of San Cristobal is 730 m (above sea level). The central highlands of Santa Cruz are also the top of a dormant shield volcano, reaching to an elevation of 864 m. The largest island, Isabela is comprised of five main shield volcanoes with a sixth, Volcan Ecuador, on its northwest tip having been partially eroded. The largest diameter shield volcano in the archipelago is Sierra Negra (see Fig. 2.12)—at 50 km—and with a maximum elevation of 1124 m., whilst the highest point in all of the Galapagos Islands is at 1707 m, on the summit of Volcan Wolf.

The typical morphology of a shield volcano has the steepest slopes where the flanks rise to the summit crater. Moving outward down the flanks, the grade of the landscape lessens. In Galapagos, the flank typically can be divided roughly into 3 sections, with 15–30° slopes near the summit, the majority of the intermediate slopes at around 5°, and with only the most distal slopes having a pitch of 1–2° (e.g. Reynolds et al. 1995). Notably, the very low slope on the distal flanks of these shield volcanoes can lead to dramatic changes in coastlines with only modest changes in sea level (see Fig. 2.12).

Fig. 2.12 Photo of Sierra Negra Volcano on Isabela Island, Galapagos. The classic shield volcano has very gradually sloping flanks rising to a summit just over 1000 masl. *Photo* K. N. Page

The elevation that is created by the shield volcanoes is also crucial to the development of ecosystem diversity. Precipitation in the highlands leads to increased soil formation and more lush vegetation and hence the highlands provide farming areas for the population of the islands. Crucially this zonation from relatively moist highland areas, often with seasonal mists, to more arid and sparsely vegetated lowland areas, is a major ecological control and many species and subspecies, including most notable tortoises, have adaptations for different climactic zones (see Chap. 3).

Listing of Notable Shield Volcanoes in Galapagos with thier elevations in meters above sea level (masl):

On Isabella Island (south to north):

- Cerro Azul—1640 masl
- Sierra Negra—largest diameter in the archipelago up to 60 km, reaching maximum elevation of 1124 masl
- Volcan Alcedo—1128 masl
- Volcan Darwin—1326 masl
- Volcan Wolf—highest point in Galapagos at 1707 masl
- Volcan Ecuador (partially eroded away)—808 masl

On other Islands:

- Highlands of Santa Cruz 864 masl
- Highlands of San Cristobal 730 masl
- Santiago Island—909 masl
- Fernandina—1476 masl

Marchena—343 masl

(b) **Calderas**—At the summit of most of the Galapagos shield volcanos is a crater known as a *caldera*. In the context of the low viscosity basaltic eruptions that typify the islands, these are depressions that are commonly the result from some collapse of a portion of the top of the volcano as magma drains out of a shallow reservoir during eruption. This collapsed depression is a caldera. Calderas can also be formed explosively. They can be formed due to the removal of the top of the volcano due to extreme explosive activity where the magma is more silicic and hence more viscous (e.g. Mt. St. Helens, in Cascades Range of Washington and Oregon in the USA). A larger caldera can also form as a result of very large volume eruption such as in Yellowstone National Park, Wyoming, U.S.

In shield volcanos collapse calderas usually have a flat floor and eruptions subsequent to collapse often come from the ring of fractures that develops between the caldera floor and the crater walls, as a result of the collapse. These craters at the summits of shield volcanoes have very steep, cliff-like sides ranging from 10's to several hundred meters in height. The shield volcanoes on the younger islands to the western side of the archipelago (Isabela and Fernandina), not surprisingly tend to have a better developed and better preserved caldera morphology. And in the east, if a caldera formerly existed in the much older shield volcano of San Cristobal Island, it is no longer preserved due to erosion.

The diameters of the caldera craters in the Galapagos ranges from a few kilometers to up to 9 km in Sierra Negra on Isabela Island (Fig. 2.13). This geomorphological structure can lead to relative isolation of populations of animals within these calderas, such as the 6,000 + tortoise population residing in the caldera of Volcan Alcedo or the pink iguanas in the caldera at Volcan Wolf (see Chapter 3).

The calderas of Galapagos are found mainly on Isabela Island, with Sierra Negra ranging up to 9 km in diameter, Volcan Alcedo, Volcan Wolf, Volcan Darwin, and Cerro Azul, as well as Volcan Cumbre on Fernandina Island.

(c) **Cinder Cones**—As molten lava and the expanding gasses in a subsurface conduit reach the surface the sudden release of gas can cause basaltic lava to be sprayed and spattered as it is erupted from a vent. As the lava is sprayed into the air, fist-sized blobs of liquid are cooled during flight and solidify before landing. These 'cinders' accumulate in a circular ring around the vent, building a small volcano known as a 'cinder cone', the product of a single eruption.

As an eruption proceeds, however, the lava will eventually become degassed and will flow effusively out of the vent without spraying through the air. Hence, most cinder cones have a small lava flow associated with them where molten lava has 'leaked' through the side of the cone, or breached part of its ring structure. There are a number of cinder cones around the Galapagos Islands including fresh cones at Volcan Chico, a parasitic cone on the flank of Sierra Negra (Fig. 2.14).

Many cinder cones become vegetated due to the more rapid break down of the porous lava to form soils and may

Fig. 2.13 View from the rim of the caldera atop Sierra Negra Volcano on Isabela Island. *Photo* D. F. Kelley

(a)

(b)

Fig. 2.14 **a** Cinder cone in the Volcan Chico eruptive area on the upper flank of the northeastern side of Sierra Negra Volcano. *Photo* D. F. Kelley; **b** well-developed Cinder Cones on the flanks of the Cerro Azul volcano, viewed from a tourist boat; Isla Isabela. *Photo* K. N. Page

Fig. 2.15 Frigate Bird Hill. This is a cinder cone that has been dissected by wave action due to its position on the shore of San Cristobal Island. It is sufficiently old that vegetation has grown over it, but the internal structure of the cone can still be seen where it has been excavated by the ocean. *Photo* D. F. Kelley

be left as *kipukas* (see below) or partially excavated by wave action where adjacent to the modern coast (see Fig. 2.15).

(d) **Tuff Cones**—Charles Darwin called tuff cones "*the most striking features in the geology of this archipelago*" (1844). They are indeed quite characteristic of the Galapagos Islands due to the high amount of active volcanism on the Galapagos platform. Tuff cones are similar to cinder cones in the formation mechanism with one important difference—the material that is ejected from the vent to build the circular cone is ash sized. As basaltic lava erupts into shallow water, the interaction between these two very different liquids causes an explosive style of eruption (i.e. a phreatomagmatic eruption). The lava is fragmented into ash-sized particles that solidify rapidly as they are ejected through the air. Thus, tuff cone formation is somewhat rare globally as it is specific to this tectonic and geographic situation (basaltic lava erupting in shallow water). Globally, tuff deposits are typically generated by thick, widespread layers of ash that are distributed regionally during very explosive, high volume eruptions of composite volcanoes (e.g. Mount St. Helens), or extreme, Yellowstone style eruptions. Darwin noted that there were no widely distributed layers of ash across the Galapagos Islands (1844). Rather, the tuffs are all concentrated around individual eruptive vents. This erupted material often includes fragments of the basaltic rock, corals, or shells that are incorporated from the sea floor through which the eruption blasted (Fig. 2.16a). If the eruption proceeds long enough, it is possible for the cone to build up above water level, such that any further erupted lava no longer makes contact with water and hence maintains a liquid basaltic lava flow that will solidify in the center of the tuff cone to a solid basaltic rock deposit. Figure 2.16b shows sheet-like dykes of basalt in Cerro Brujo (Witch Hill), a tuff cone that has been dissected by wave action as the cone is on the coast of San Cristobal Island. These dykes are solidified lava that flowed into cracks in the structure of the Tuff Cone towards the end of the eruptive event.

There are scores of tuff cones around the Galapagos archipelago as the combination of basaltic volcanism and an oceanic platform that often provides an eruptive setting in shallow water—each representing a single eruption. Most of the tuff cones which remain isolated in the sea off the coasts of the islands are typically heavily eroded, creating visually striking landforms. A well-known example is Leon Dormido (Kicker Rock) (Fig. 2.17a), a popular snorkeling and diving site 5 km off the west coast of San Cristobal Island.

The erosion of tuff cones or wider tuff rings most commonly occurs on its south side due to the prevailing force of wind and current around the Galapagos Archipelago. Tortuga Island near the southern coast of Isabela is the largest example of the resulting 'C'-shaped island, opening to the south (Fig. 2.18).

Fig. 2.16 Cerro Brujo, San Cristobal Island; **a** close up view of the texture and composition of the rock. It is seen to be fragmented and contains clasts of the seafloor through which the eruption occurred. The white clast is carbonate, while the black is basalt. Pen for scale. *Photo* Sheridan Ackiss; **b** the dissected tuff cone with internal dikes exposed. *Photo* D. F. Kelley **c** well-bedded water-lain volcanic ash in a low cliff at Puerto Egas, Isla Santiago. *Photo* K. N. Page; close up of tuffaceous material—note small fragments of shattered lava. *Photo* D. F. Kelley

Other tuff cones in Galapagos, originally formed in near coastal shallow waters, have since been 'annexed' by the growth of the adjacent island through lava flows extending the coast, such as Cerro Brujo, discussed above (Fig. 2.16). Other tuff cones have been 'assimilated' in this way now lie some distance inland from the coast as can be seen on the eastern distal flanks of Sierra Negra Volcano on Isabela Island. There are scores more examples across the platform. Many of the islets or rocks jutting out of the surf around the archipelago are the eroded remains of tuff cones.

(e) **Kipuka**—A kipuka is an area of land surrounded by a younger lava flow or multiple flows. This is a topographic high, such as a small hill or cinder cone. Because basaltic lava is quite fluid, it will flow readily into areas of lower elevation, and will not cover areas of higher elevation, especially these hills. Kipuka are widespread in the Galapagos and vary in size from a few 10 s of meters to much larger features. As a basaltic landscape ages, soil and vegetation develops, then, when a later eruption occurs, fresh, black basaltic rock is created. Therefore, kipuka can often be seen as green hills sitting in a black lava field such as those in Fig. 2.19b on the flanks of Sierra Negra volcano on Isabela Island.

2.2.5.3 Lava Flows and Features

Much of the lava from the eruptions of the volcanoes of Galapagos simply flowed as sheets and as channelized rivers across the landscape and down the gently sloping flanks of the volcanoes before eventually freezing. These flows 'repaved' the landscape with each new eruption, filling in any pre-existing topographic features such as valleys. The resulting basaltic rock is black in color, creating a barren landscape, which is truly one of the most notable and widespread features that characterizes the terrain of basaltic islands. Such landscapes are characteristic of areas of the Galapagos around the most recently erupted volcanoes, including for example the basalt field inland from Puerto

Fig. 2.17 Leon Dormido or Kicker Rock. *Photo* D. F. Kelley. This tuff cone has been eroded on all sides by wave action as a result of its location standing alone in the sea 5 km off shore on the west side of San Cristobal Island. The lower figure is the sketch of Darwin (1844)

Villamil on the island of Isabela, covering the area around Cerro Pelado on the distal flanks of Sierra Negra volcano (Fig. 2.20).

Another dramatic example of a landscape covered with lava flows is associated with Voclan Chico, an eruptive center on the high flank of Sierra Negra volcano near the caldera rim (Fig. 2.21).

In the 'older' islands such as Santa Cruz and especially San Cristobal, although such landscapes would have once existed, many of the lavas flows are weathered and vegetated (Fig. 2.22).

Fig. 2.18 Satellite imagery of Tortuga Island. The south east coast of Isabela Island and the town of Puerto Villamil are several kilometers to the northwest of Tortuga

Lava flows demonstrate a range of surface and internal features, including discrete volcanigenic landforms, the most characteristic of which can be classified as follows:

(a) **Pahoehoe** lava is created by the hottest and hence least viscous lava that whilst flowing develops a thin crust that is deformed by the flowing liquid beneath it. The crust is then 'bunched up' or folded and freezes with a ropey, but essentially smoothed texture (Fig. 2.23).

(b) **AA**—In contrast to *pahoehoe*, aa lava develops when the basaltic lava is cooler and more viscous, perhaps near the end of a large flow. As a result, the brittle, cooled surface of the lava breaks up to form a blocky, sharp and very rubbly appearance (Fig. 2.24). These landscapes can be very difficult to traverse.

(c) **Pillow lavas**—Where fluid basaltic lava erupts under water from submarine vents the lava rapidly forms a crust as it is cooled by the sea water. As lava continues coming from the vent, the growth of an individual lobe is prohibited by the solidified rim that has formed, and another lobe of contained lava will spill out alongside, rapidly forming its own solid rind. As the lava continues erupting, it forms a pile of these round, but somewhat flattened lobes that resemble a pile of pillows. The term 'pillow-lava' perfectly describes the resulting lava deposit. These deposits often form at some point during tuff cone formation as lava is erupting in a submarine setting. The best exposure of pillow lavas in Galapagos is on South Plazas Island, a small islet just off of the eastern coast of Santa Cruz. This island is a heavily eroded portion of a tuff cone with ash deposits and pillow lavas exposed through erosion. Other exposures of pillow lavas across the Galapagos are uncommon as the ongoing subsistence of the islands tends to take such features into deeper water rather than uplifting them above sea-level.

(d) **Tumuli**—As fluid basaltic lava is flowing over a very gently sloping terrain, or is ponded in a topographically low area, the surface of the liquid will solidify first, as it is in contact with the cooler air above, and will form a 'crust' on the top. As the fluid lava continues to flow beneath this crust, it can cause local inflation, which cracks the plates of solid crust and buckles it upward forming a mound. These tumuli can be preserved as localized upward bulges of plates of the surface of the lava flow (Fig. 2.25).

(e) **Squeeze-up structures**—In a situation such as that described above for the formation of tumuli, ponded liquid basaltic lava can often squeeze up through the cracks formed between the surface plates of solid rock. This degassed, thick lava can ooze up forming bulbous mounds, or linear patterns along the length of a crack (Fig. 2.26).

(f) **Hornito**—If the lava continues to squeeze out from one place due to the overburden of the solidified plates, a small rootless cone can develop around this eruptive center. The cone may have many oozing, dripping layers that are

Fig. 2.19 **a** General view and detail of a kipuka on the west coast of Isla Santiago, opposite Sombrero Chino ('Chinese Hat')—note 'island' of older, weathered and now brownish in colour and vegetated lava flow, surrounded by black, younger flow (viewed from tourist boat visiting Sombrero Chino). *Photo* K. N. Page; **b** Kipukas can be seen in the middle distance toward the right of the photo. East flank of Sierra Negra Volcano, Isabela Island. *Photo* D. F. Kelley

Fig. 2.20 Basaltic lava flow in the area of Cerro Pelado on Islabela Island outside of the town of Puerto Villamil. The relatively young basaltic rock is mostly free of vegetation. *Photo* D. F. Kelley

preserved as the lava freezes. The small rootless cone can be called an hornito and could consist of some spattered lava as well if the gas content was sufficiently high (Fig. 2.27).

(g) **Lava Tunnels**—When basaltic shield volcanoes erupt, the lava flowing from the summit can travel great distances; in many instances down to the shores of an island. This is in part due to the low viscosity of the hot basaltic lava, but also a result of the way flow systems can develop. As large volumes of lava erupt and begin to flow down the flanks of the volcano, the liquid tends to channelize following lower areas in the topography. As lava begins to flow in channels, a crust will form on the top where the lava is more prone to freeze in contact with the relatively cold air. This solid rock roof now insulates the flowing channel of lava below from the cooling effect of the air and hence an established tunnel allows lava to flow great distances down the flank because it remains hot and molten. Branching channels of lava tunnels can be established all the way from the summit to the coast at base of large shield volcanoes. When the eruption ceases, the last of the lava flows down as far as it can before freezing, leaving many drained lava tunnels in the highlands (Fig. 2.28). Indeed, some lava tunnels can be used repeatedly by multiple pulses of one eruption or by lava from a series of eruptions. Lava tunnels can be up to 15 m in height and are a popular feature for tourists to explore, particularly on the island of Santa Cruz where they are quite abundant. As a portion of an island with lava tunnels ages, however, the tunnels tend to collapse as the physical and chemical weathering of the rock compromises the integrity of the roof.

In some places around the coasts of the islands, lava tunnels that reached the sea are being eroded by wave action. In such cases, arches of rock can be left behind creating interesting landscape both above and below water. Cabo Rosa on the southern shore of Santa Isabel is one such place where tour groups are taken for photos of the arches with cactuses, and snorkeling amongst the pillars that can be seen under the water (Fig. 2.29).

Fig. 2.21 Basaltic Landscape at Volcan Chico area on the flank of Sierra Negra Volcano, Isabela Island. *Photo* D. F. Kelley

2.2.5.4 Lava Petrology (i.e. Chemistry and Mineralogy)

As the volcanic rocks of the Galapagos Islands are entirely of a basaltic mineralogical composition, having formed through the solidification of a basaltic lava either effusively or explosively, there is not a tremendous amount of diversity in the mineralogy of the rocks; that is, the assemblage of different minerals of which the rock is composed. An igneous rock is defined as basaltic if it has between 45 and 55% SiO_2 by weight when analyzed for its total chemical composition. When a given chemical composition of lava solidifies, it does so through the forming of crystals of certain minerals. Different lava compositions will crystallize different assemblages and abundances of minerals because of the percentages of the atoms of different elements that are available to bond to one another to create the mineral crystals. Basalt rock is generally composed of the minerals olivine, pyroxene (perhaps coexisting orthopyroxene and clinopyroxene), and plagioclase feldspar (see Fig. 2.30). The rocks of Galapagos are no exception.

However, these islands are a classic example of basaltic ocean islands created over a mantle plume, and so they are one of a handful of localities on Earth where basalt from this type of source can be described. Thus, in Darwin's visit to the islands in 1832, he made detailed observations of the mineralogy that can be observed in hand specimens. This latter point is of import because volcanic rocks are generally very fine grained, due to the very rapid cooling of the lava such that the crystals cannot be identified by the naked eye or even without the use of a petrographic microscope (note the scale bars in Fig. 2.30). In Darwin's time, the naming and classification of minerals was different than it is today, as techniques such as X-ray diffraction and electron microscopes were not available. However, a range of physical tests, including optical, were available and the relied virtually entirely on mineralogy. Darwin carefully noted the presence, abundance and size of olivine ($(Mg,Fe)_2SiO_4$), augite ($(Mg,Fe,Ca)SiO_3$), albite ($NaAlSi_3O_8$),as well as bubble-like *vesicles* that formed as lava rapidly solidified trapping bubbles of gas.

Darwin noted that dark crystals of olivine and augite were observed in the pale tuff deposits of San Cristobal Island (1844, p. 100) (see Fig. 2.16a), the typically very small dark-colored crystals were difficult to distinguish in solid basaltic rock. As the lava was pulverized during the phreatomagmatic eruption that creates a tuff cone, some of the mineral crystals that formed during the magma's ascent towards the surface—hence with longer to form, they can be slightly larger than those in the rest of the groundmass of the rock—are sent flying through the air along with the ash particles to be included in the layers of tuff. Also in the tuff are clasts—i.e. inclusions—of marine rock such as fragments of coral or calcareous sand and some of the older, solid basaltic rock that formed the oceanic crust though which the plume has passed. These observations helped Darwin to understand the mechanism of eruption and formation of the tuff cones (Fig. 2.16).

Fig. 2.22 Entrance to a lava tunnel named "Tuneles de Sucre" within a heavily vegetated lave flow on Isabela Island. *Photo* D. F. Kelley

Darwin also described the presence of mineral crystals in the lava flows of the islands on which he made landfall. He described the varying amounts of olivine, augite, and albite on Isabela, Santiago, and San Cristobal. However, it was the size, apparent fluidity, and surface structures of the lava flows that he put the most attention toward, as there is a relatively homogenous mineralogy throughout these basaltic islands.

2.2.5.5 Eruption Types and Chronologies

Although the products of eruptions such as lava, ash, cinder cones, and the volcanoes, together with their relative chronologies, are clear features of the geodiversity of the islands, active eruptions themselves are also a facet, analogous to active fluvial or coastal systems. As this is a volcanically active area, the volcanoes, particularly of the western end of the archipelago, continue to erupt periodically. These eruptions can be spectacular to behold, often including lava fountains, 'curtains of fire', and long branching lava flows moving down the flanks of the volcanoes—all of which are incandescent at night. The Galapagos Islands are, therefore, important for advancing scientific understanding of these processes and the sampling and analysis of rocks created during these eruptions advances our understanding of the generation, migration, storage and eruption processes of the magma/lava.

While tourists do not generally visit the Galapagos Islands with the expectation of seeing erupting lava as they might in Hawaii, eruptions certainly provide an unexpected bonus to some lucky visitors and can foster geological interest and enthusiasm. Tour vessels often divert from their planned itineraries in order to provide this experience to their guests in the event of an eruption.

The populated areas of the Galapagos Islands are relatively few and far between, due to the difficult terrain, relatively short history of settlement, and more recently the restrictions imposed by the development and expansion of the Galapagos National Park. As a result, the geological hazards associated with the eruptive activity of the islands is relatively low. The settlements on the island of Isabela are

(a)

(b)

Fig. 2.23 'Ropey' appearance of the surface of a *pahoehoe* lava flow; **a** deposits on the beach at Puerto Villamil, Isabela Island. *Photo* D. F. Kelley; **b** Punta Moreno, Isla Isabela (with the flanks of the volcanoes Sierra Negra and Alcedo just visible on the skyline to the right and left respectively). *Photo* D. F. Kelley

(a)

(b)

Fig. 2.24 Typical blocky, angular appearance of the surface of an a'a lava flow. Cerro Pelado lava field, Isabela Island. *Photo* D. F. Kelley

Fig. 2.25 Tumuli structures on San Cristobal Island. The plates of the surface of a basaltic lava flow were buckled upward by pressure of the flowing liquid forming the cracks that are seen on the surface. *Photo* D. F. Kelley

Fig. 2.26 Squeezed up lava that has extruded from the crack in a pahoehoe surface of a freezing lava flow. Punta Picuna, San Cristobal Island. *Photo* D. F. Kelley

Fig. 2.27 Layers of lava have dripped on top of each other while squeezing out of a crack on the surface of a lava flow crust forming this small hornito. The card in the photo is 15 cm in length. Punta Picuna, San Cristobal Island. *Photo* D. F. Kelley

more at risk than those on Santa Cruz, San Cristobal, or Floreana simply because of their location near the west end of the archipelago, above the hotspot, and more specifically their location on the active Sierra Negra Volcano. Remarkably, however, there appear to be no examples in the history of settled populations on Galapagos of loss of life due to the proximity of the eruption to human habitation.

In this context, however, it is the geological record of successive eruptions and the style of those eruptions—either observed or inferred from deposit types—that forms a key facet of the island's geodiversity. Although all volcanic landforms and deposits, including of ash and lava, record phases of activity, those documented and studied, plus those showing, for instance the oldest deposits on each island, or records of exceptional events (e.g. size, explosivity, etc.) can have a particular scientific and historical significance.

2.2.5.6 Geothermal Springs and Fumaroles

Geothermal heat driving hot springs and fumaroles are associated with all volcanic regions and although most tourists do not see them, the Galapagos are no exception. The periodic volcanic activity on the islands is fed by magma which is stored in the crust at depths as shallow as 3 km, for instance below Sierra Negra (Reynolds et al. 1995). At around 1,200 °C, this magma provides a lot of heat to the shallow crust and as a result, for instance around Sierra Negra, there are *fumaroles* (vents of steam and gasses) and deposits of sulfur minerals associated with degassing (Fig. 2.32). The deposits of sulfur in an area on the southwestern edge of the rim of the crater atop Sierra Negra Volcano on Isabela Island referred to as the Sulfur Mines. While there is no actual history of mining there, the name is given to this site, which is the destination of a guided hiking/horseback riding trail available to visitors of Isabela Island (See Chap. 5) (Fig. 2.31).

The geothermal heat emanating from the crater floor in Alcedo Volcano on Isabela Island provides a unique opportunity for the tortoise population that lives in the caldera. These tortoises, *Geochelone nigra vandenburghi* (Desola) do not have to migrate to the coastal areas to find the warm environment necessary to lay their eggs as do the other highland populations of tortoises across the Galapagos Islands. Instead, they lay their eggs along the coast of the lake that exists within the caldera. Due to the geothermal energy of the volcano, the ambient temperature is sufficient for incubation. Due to this behavior, this tortoise population was not discovered and described until 1930, and had not been hunted to endangerment by early human visitors to the islands as many others have. The population is now over 6000, one of the largest in Galapagos (Beheregaray et al. 2003).

2.2.5.7 Soils

With time, basaltic rock, as most others, will chemically weather due to interaction with rain and the atmosphere.

Fig. 2.28 Lava tunnel "El Mirador" just outside of Puerto Ayora on Santa Cruz Island. Smooth sides can be seen in the foreground on the right. On the left in the middle distance is a shelf where the level of the lava flowed intermittently in a subsequent use of this tunnel. A skylight has formed by collapse of part of the roof of the tunnel. *Photo* D. F. Kelley

(a)

(b)

Fig. 2.29 Eroded coastal lava tunnels, now forming natural arches. **a** Cabo Rosa, Isabela Island. *Photo* D. F. Kelley; **b** near Puerto Egas, Isla Santiago. *Photo* K. N. Page

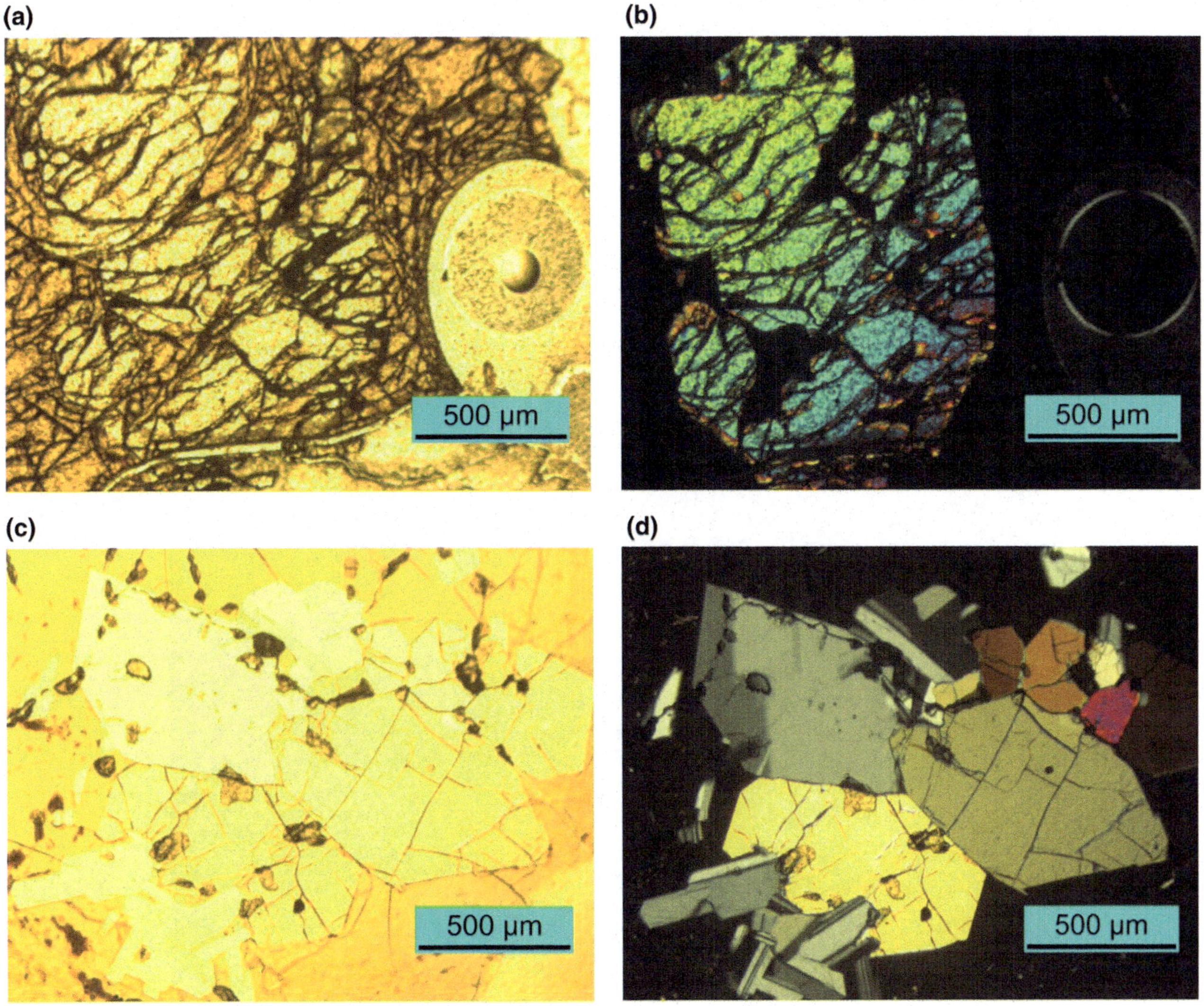

Fig. 2.30 Thin section of basalt from Kalfstindar, Iceland, showing the characteristic mineralogy and texture these volcanic rocks. Thin sections are produced by grinding and polishing thin slices of rock attached to a microscope slide down to 30 μm in thickness in order to be viewed by using a petrographic microscope which passes light through the section from below toward the eye. The photos in panels (a) and (c) are viewed with regular white light passing through the thin section. Those in panels (b) and (d) are shown with polarized light passing through the thin section. Polarized light (i.e. light waves all oriented in the same direction) is then reoriented and/or slowed in different ways as it passes through different minerals resulting in characteristic *interference colors* allowing for mineral identification. In Panel (b), an olivine crystal is shown with irregular fractures and yellow, blue, violet interference colors. In Panel (d), a cluster of plagioclase feldspar (striped in shades of grey and black) and pyroxene (larger crystals with tans and greys in the center of the cluster) is shown

Initially the minerals will alter and so the rock surface takes on a greyer appearance, but ultimately the oxidation of the dominant iron-rich minerals will lead to the formation of soils and subsoils strongly colored by red and yellow iron-oxides. The minerals formed due to this chemical weathering are also softer than the original rock and so can be washed away, hence contributing to the physically breakup of the rock. The accumulation of these alteration minerals is the first step in soil development upon a new lava field. As the first plants take root, they contribute to the physical breakdown of the rock, and when they die, they both contribute organic material to the development of soil as well as releasing organic acids that promote further breakdown. Around the Galapagos Islands, the relative ages of lava flows can, therefore be estimated based on the color and amount of vegetation.

Due to the nature of eruption of basaltic lava, however, there have not been any eruptions that might cover all of the islands with layers of ash, unlike in the Andes, for example, where thick ash deposits lead to a rich soil development. As Charles Darwin noted, "*Owing to the absence of ashes, and the general indecomposable character of the lava in this archipelago, the islands are slowly clothed with poor vegetation, and the scenery has a desolate and frightful aspect*". [1851, p. 62].

Fig. 2.31 Sulfur-bearing minerals deposited around a small fumerole vent in the area of Volcan Chico on the flank of Sierra Negra Volcano, Isabela Island. The heat coming from this vent can be noticeably felt with one's hand. *Photo* D. F. Kelley

Although the dominance of basaltic substrate means that the range of soil types in the Galapagos is limited, there can still be significant differences in thickness and maturity, depending on how long the soil has had to develop and the altitude-controlled humidity in which it has developed.

In Galapagos, soils are better developed on the older islands, and within islands are better developed in the highlands where rain is more common. Therefore, in the highlands of Santa Cruz, the soils are ~ 1 m in thickness, and in the highlands of San Cristobal, the soil is up to 2 m thick (see Fig. 2.33).

Although soil types are a major control on vegetation and agricultural potential, they are also a facet of geodiversity, especially where process-related classification schemes have been developed—and hence the type localities for different types may require conservation so that they can continue to provide a reference for ongoing research, including agro-environmental.

Fig. 2.32 The Sulfur Mines site at Isabela Island. Steam can be seen rising from fumeroles near the top of the hill. The white colored slope is covered with sulfur bearing minerals that have been deposited by chemical rich waters condensing at the fumeroles and flowing down the slopes of the hillside. *Photo* D. F. Kelley

2.2.5.8 Fluvial Geomorphology and Hydrogeology

Although the majority of landform features across the Galapagos are volcanically generated, valley forms do exist where minor streams—often seasonal have eroded into bed-rock and superficial deposits, including soils. The only permanent freshwater lake as well as the only permanently flowing stream area on San Cristobal Island on the older, southwestern portion of the island—this portion of this island is the oldest in Galapagos. Floreana, Isabela, and Santa Cruz islands also have small streams and pools, but they are temporary and dry up during drought, with the latter often becoming brackish (i.e. salty), as on Isabela. Locally springs and pools—often seasonal, developing during the wet-season can form oases for plants and animals, including tortoises (Fig. 2.34).

Some water bodies, such as caldera lakes may also develop and persist for several years during climatic events such as El Niño, before drying up, for instance in Cerro Azul on Isabella (d'Ozouville 2009). Currently, however, there appears to be no systematic landscape survey related to hydrological and fluvial features in the Galapagos.

The search for fresh water had been a primary goal of Captain Fitzroy during the Beagle's visit to the Galapagos Islands in 1835 as supplies needed to be replenished before travelling onward to Tahiti (Grant and Estes 2009). On their fifth day in the islands, they found the needed fresh water on the southeast side of San Cristobal at a location now called Freshwater Bay (Bahia de Agua Dulce)—where the only permanent stream reaches the sea.

2.2.5.9 Coastal Geomorphology and Marine Deposits

The archipelago character of the Galapagos means that coastlines are very extensive and hence a very wide range of features are present, including hard and soft rock cliffs,

Fig. 2.33 On this hiking trail around Junco Lagoon atop the shield volcano of the southwestern portion of San Cristobal Island, there is 1–2 m of soil developed. *Photo* D. F. Kelley

Fig. 2.34 Tortoises in spring in the El Chato 2 reserve—a private touristic venue, Santa Cruz. *Photo* K. N. Page

beaches, barrier beaches, brackish and saline lagoons, shore platforms, arches, stacks, blow-holes, etc., etc. Many represent erosional features developed as volcanic landforms are eroded, but some, such as the barrier beaches and lagoons have been formed in coastal deposition regimes (Fig. 2.35).

Some of the most important coastal features related to the volcanic Geoheritage are raised marine platforms, some with 'beach' deposits. Darwin probably first noted the presence of elevated marine sedimentary deposits with modern shells on San Cristobal island in 1835 (Darwin 1844). There are similar locations around the islands, including most notably Bahia Urbina on the west side of Isabela Island.

2.2.5.10 Palaeontological Heritage and 'Cave' Deposits

Although the *living* natural heritage of the Galapagos is globally famous, the Galapagos also has a rich palaeontological heritage, and not just of recently and anthropologically extinct species. This heritage includes not only the marine invertebrates of the raised marine deposits mentioned above—as already observed by Darwin on San Cristobal (Nicholls 2014)—it also includes important 'cave' deposits rich in vertebrate remains, especially 'microvertebrates', in some of the lava tunnels. As reviewed by Steadman (2009), such deposits take the history of the Galapagos fauna back to nearly 22,000 years into the Pleistocene—although most sites are of Holocene age (i.e. less than 10,000 years old) and some definitively of 'historical' age.

Many of the microinvertebrate faunas have been recovered from fissure and lava cave deposits, the latter often concentrated by owls feeding nestlings within these systems. The faunas include rodents (in particular the only endemic group, rice rats of the genus *Nesoryzomys* and one species of *Aegialomys*), birds, tortoises, snakes and other reptiles—some of which are extant, others extinct. Preservation appears to be best in more arid terrains across the islands, as tropical, humid conditions lead to relatively rapid decay of bone material and faunas have been recovered from San Cristobal, Santa Cruz, Floreana, Rabida and Isabela. Steadman (loc. Cit.) recorded 15 microvertebrate sites across these islands but many more are likely to remain undiscovered or unobserved, hence any attempt at listing must be considered 'open'.

2.2.5.11 Moveable Galapagean Geodiversity

As research on the geology of the Galapagos has progressed, samples of rocks, minerals, sediments and fossil specimens

Fig. 2.35 Coastal features. **a** Coastal barrier-beach-lagoon system, near Playa Roja, Isla Rabida. **b** sea-stack, 'Buccaneer Cove', Isla Santiago. **c** coastal arches developed in well-bedded water-lain volcanic ash (='tuff') Sea-stack, Puerto Egas, Isla Santiago. **d** sand-beach system, Playa Espumilla, Isla Santiago. *Photos* K. N. Page

will have accompanied researchers back to their institutions. Although modern research permits are likely to specify a national repository for the safe guard for the future of key specimens of particular scientific interest, in the past a significant amount of material will have left the islands with little or no itemized record.

In the case of most rock samples, however, there is no significant conservation issue associated with the collection of common rock types, providing that good examples of associated features, e.g. pahoehoe surfaces and sensitive ecological features, are not permanently damaged or disturbed. Such specimens fall into 'Category 4' of Page (2004, 2018). i.e.

"*Common and representative species and specimens, well represented in nationalmuseums and other institutions, or sufficiently abundant that any non-scientific collecting or removal will not prejudice future scientific work; also includes specimens collected loose, for instance from scree, rubble or beach material, where the lack of stratigraphical information significantly reduces scientific use*". In the case of such material, providing that the operation of sampling is approved by the GNP and carried out within all applicable statutes and guidelines, there is absolute no need that the eventually repository for the specimens be dictated. Indeed, representative and legally collected geological specimens in institutions across the world can only raise awareness and stimulate future studies, which could greatly benefit the ongoing scientific program of the CDRC and the GNP.

Fossil material in particular, however, especially where it is rare or restricted to very limited deposits such as caves, however, requires quite a different approach and categories 1 and 2 might apply, e.g.:

Category 1: "Type (a), figured (b) and cited (c) specimens: The first (Category 1a) are fundamental to the definition of fossil species as regulated by the International Commission on Zoological Nomenclature (a UNESCO project); the latter two categories (1b and 1c, respectively) underpin all palaeontological studies as supporting material or evidence of scientific observations or conclusions"....

"*Legal systems should on the one hand ensure that...access [for research] can take place and on the other hand seek to guarantee that institutional deposition and full protection of the relevant described specimens is achieved once study is completed*".

Category 2: "*Unique, rare or exceptionally complete or well-preserved taxa or specimens or assemblages of specimens of fundamental importance to actual or future scientific studies. Category 2 specimens are crucial to the science of palaeontology, as the raw material for ongoing or future studies. Conservation and legal systems or practice should, therefore, ensure (including through the use of expert advisors or assessors) that such specimens are deposited and protected within nationally recognized institutions, where they will remain accessible for future study and appreciation.*"

Today, it would be expected that the deposition of such specimens within a regional or national institution, for instance a national natural history museum, would ultimately take place—for instance after research had been completed in a foreign or even another national institution, such as a University. However, such stipulations are not always applied and certainly not historically. As a result Galopegean geological heritage, now has a wide geographical distribution.

The oldest and most famous of Galapogean geological collections is of course that Charles Darwin collected himself and which now resides in the Sedgwick Museum of the University of Cambridge, England (www.sedgwickmuseum.org). Part of this collection—including his hammer—has recently been placed in a new display in the museum, entitled '*Darwin the Geologist*' as part of the bicentenary celebrations of Darwin's birth, including an online resource at http://www.sedgwickmuseum.org/index.php?page=darwin (accessed 3/2018).

Samples of basaltic lava from the Galapagos are included in the displays and form part of the museum's 'Beagle' Collection, which comprises around 2000 rocks and a few fossils collected by Darwin himself during his voyage on H. M. S. Beagle from 1831 to 1836 (Porter 1985; Desmond and Moore 1991). The collection was given to the Museum after Darwin's death and manuscript catalogue prepared by the famous Cambridge petrologist, Alfred Harker (1859–1939), including entries from Darwin's notebooks (http://www.sedgwickmuseum.org/index.php?page=the-beagle-collection).

Although Darwin's collection from the Galapagos is relatively small, it is undoubtedly the most famous from the islands. There are, however, representative suites of specimens and derived data held in many other institutions across the world. Crucially, and especially where data and specimens might be available to other specialists, this resource can help inform future studies.

As discussed above, however, there may be cases where modern approaches to science and conservation would favor the return—or repatriation—of key and unique specimens to Ecuador as part of a national natural heritage, but such processes, however well-intentioned, can be complex, to say the least. Lima and Ponciano (2017) document such a case in Brazil, where an important collection of Devonian fossils was eventually returned from the USA, but this had always been the intention of the researchers. Commonly, however, there is a great reluctance, even refusal to return such materials, as the conservation agency of the Government of Scotland, Scottish National Heritage, discovered when it tried to reclaim stolen Scottish fossil material from a museum in Germany (Macfadyen 2006).

However, where national institutions can offer the same level of environmental control (including museum conservation standards) and security that any elsewhere can, there is no longer any justification for institutions in other countries to continue to covert important specimens from elsewhere. But this is a global problem and despite agreements such as UNESCO *Convention on the Means of Prohibiting the Illicit Import, Export and Transfer of Ownership of Cultural Property* (1970; http://portal.unesco.org/en/ev.php-URL_ID=13039&URL_DO=DO_TOPIC&URL_SECTION=201.html), there has often been little real action. The infamous so-called 'Elgin marbles', part of a carved marble freize *privately* removed from the ancient Greek, Parthenon in Athens by a Scottish aristocrat in 1801—and then sold to the British government—is a 'classic' even 'classical' example. Although now housed in the British Museum in London in one 'developed' European country, the institution still refuses to return them to Greece, another 'developed' European country. Interestingly, the justifications the Museum uses to keep the marble freizes in England (http://www.britishmuseum.org/about_us/news_and_press/statements/parthenon_sculptures.aspx) have a similar neo-colonialist tone to some of the arguments used in the establishment of National Parks across the world to a Euro-North American model (see discussion in Chap. 4—there is also an informative discussion on the 'Elgin Marbles' at: https://en.wikipedia.org/wiki/Elgin_Marbles).

Fortunately, however, as most Galapogean geological specimens are samples of the vast expanse of volcanic rocks that form the islands, they do not fall into a similar category of 'Unique' or 'Internationally important' (see Sect. 2.3 below), but are simply 'Representative' of the islands geological history. Fossils, as ever are more problematic (see discussions of Page 2018), but some 'grading' of relative scientific value is always possible. Crucially, the necessity for ongoing conservation work and storage under stable environmental conditions to minimize deterioration, may make the choice of eventual repository for future study and generations much more restricted. But things do change, and where once there might not have been funding for modern environmentally controlled collections storage, such facilities are becoming much more common globally.

2.2.5.12 Geoheritage in a Cultural Context

Although most of the modern construction of the Galapagos utilizes imported cement and concrete, lava rock is used for some buildings, roads, parking areas, and decorative stone. There is a quarry on each inhabited island where active excavation is ongoing for the purpose of supplying raw materials for the ever increasing development of the inhabited portions of the islands. Where roads were once all dirt or sand surfaces, they are being repaved with a mixture of imported tar from the mainland and crushed rock from the local quarries. This includes also development of the airports, shipping ports, etc. The quarries are generally excavating the vesicular, loose, basaltic scoria material of existing cinder cones. On the island of San Cristobal, there is a quarry just outside of Puerto Baquerizo Moreno (Fig. 2.31). On the most populated island of Santa Cruz, there are two such quarries—Mina de Granillo Negro and Mina de Granillo Rojo. As the names suggest, one is mining black colored scoria and the other red. The cinder cones that are mined of course have a finite amount of available material and Fig. 2.36 demonstrates that for the cone being actively quarried on San Cristobal, only around half of its original structure and deposits remain.

On Isabela, there was formerly a quarry site to supply material for use by the municipality at El Chapin. Although, virtually all accessible and useable material has now been removed from the site, the brackish lagoon which subsequently formed in the excavation is now a popular feeding site for flamingos—and thus a popular stop for visitors to stop and take photos (Fig. 2.37).

Once the El Chapin quarry was exhausted, the municipality of Puerto Villamil began mining operations at Cerro Pelado (Fig. 2.38), just outside of the town. Material is crushed and sorted by size on site as can be seen in Fig. 2.38.

The basaltic scoria is not the only material that is useful for such practical purposes. In order to provide a rudimentary cementation, crushed basaltic rock is often mixed with the calcareous sand that can be mined from elevated deposits of former beaches. One such "perched" beach exists on the property of the airport outside of Puerto Villamil on Isabela Island (Fig. 2.34). This locality is now $\sim$5 m above sea level and nearly 2 km inland from the coast having been isolated from the sea by an inflow of lava from an eruption of Sierra Negra volcano that extended the size of the island within the past few thousand years (Fig. 2.39).

Working and disused sites such as those described above not only demonstrate a societal interaction with a geological heritage, which may have some cultural significance in their own right, they also provide excellent opportunities for viewing and studying geological materials beyond the restrictions imposed on sites with the GNP. Crucially, however, rather than always being a 'threat' to the iconic ecology of the islands, with sensitive restoration, or simply non-intervention, they provide additional resources and

Fig. 2.36 Municipal quarry outside of Puerto Baquerizo Moreno. *Photo* D. F. Kelley

Fig. 2.37 El Chapin, an inactive quarry near Puerto Villamil that is now filled in with a brackish lagoon. *Photo* D. F. Kelley

habitats for the islands wildlife, the presence of Carribean Flamingos (*Phoenicopterus ruber* Linn.) and other bird species in brackish pools being an excellent example.

In addition to the use of tephra within constructional material and as a surface dressing, blocks and slabs of basaltic lava is also often used for decorative purposes around the towns of all of the islands. Cut slabs of the vesicular basalt are used on sidewalks, benches, and low walls around the public spaces of the towns. The rougher, natural slabs of the tops of lava flows are often used in this same way on the sides of buildings or public areas (see Fig. 2.40). This material is by its nature quite strong and resistant to weathering, but also provides an aesthetic that connects to the geologic nature of the islands.

2.3 Inventories and Selection Procedures for Identifying a Key Geoheritage

2.3.1 Introduction

Even within a highly protected area such as the Galápagos National Park, the identifications of key features to inform conservation practice is essential. Most obviously, this includes endemic species, and systematic studies will help identify those most in need of interventive management, including captive breeding as for certain tortoise species (for instance at the Arnaldo Tupiza Giant Tortoise Breeding Center on Santa Isabella), habitat recreation or the elimination of invasive species, including parasites (as in the case of those afflicting the Mangrove Finch—*Camarhynchus heliobates*). In addition, other sensitive areas such as prime habitat and breeding sites will also need to be identified, to inform site usage by both visitors and residents. Inevitably, some of the protective measures implemented can be controversial, but should ideally be supported by rigorous and objective scientific justifications, rather than just subjective 'opinion'.

Similarly, geological features require some selective framework through which to identify key or unique features, not only to inform specific management actions, but also to reduce the risk of inadvertent loss or damage due to implementation of other management actions—for instance the establishment of access routes or habitat management.

2.3.2 Inventory—Defining the Resource to Inform Management

The two contrasting approaches to selecting 'key' features for conservation—or in other words establishing an 'inventory'—are either through an 'expert' system based on the opinion of appropriate specialists on the types of geological features and processes present or to develop some form of 'scoring' system, through which to 'grade' features and areas. In practice, however, the establishment of criteria, for instance for identifying international, national, regional and

Fig. 2.38 Quarry at Cerro Pelado outside of Puerto Villamil on Isabela Island. *Photo* D. F. Kelley

local importance, or classifying the features present (e.g. 'cinder cones', 'tuff cones', 'hornitos', etc.) are essential to both approaches.

The 'expert witness' approach to site selection is exemplified by the Geological Conservation review programme established for England, Scotland and Wales in 1977 by the then Nature Conservancy Council, to provide a rigorous and systematic approach to site selection as a basis for legal designations as protected sites.

The project employed a wide range of geological specialists, ultimately selecting around 3,000 sites of at least national geological and geomorphological importance, within 97 subject 'blocks', representing key geological themes such as subdivisions of geological time and process-related categories (http://jncc.defra.gov.uk/page-2947). Although no specific criteria were listed at the time, within each geological or geomorphological theme, sites were effectively selected for three basic reasons (Ellis et al. 1996):

1. *Sites of international importance* (e.g. international stratigraphical reference sections such 'GSSPs' (Global Stratotype Section and Point), type localities for important fossil species, sites with exceptionally well-preserved fossil assemblages, rare or unique mineral deposits, etc.).
2. *Exceptional* sites (e.g. with unique, rare or unusual features, also 'Highlights' of British geology).
3. *Representative* sites (e.g. showing characteristic features of the GCR theme, for instance representing the main rock units of that age in the UK, or good examples of specific landforms or processes).

The effectiveness of such an approach, inevitably depends on the experience of the thematic specialist involved, and the success of the GCR process can be recognized 40 years later, as relatively few significant omissions of nationally important sites have subsequently been

Fig. 2.39 Deposits of layers of calcareous sand that were formed when this area was at the shore of Isabela Island and consisted of a beach. This location at the Puerto Villamil Airport is now quarried for the calcareous material. *Photo* D. F. Kelley

identified, although the exact relationship between nationally selected GCR sites and later regional site networks remains to be established (Page 2016).

Due to the high resource implications of such a process, few other countries have attempted a review of their geological heritage as comprehensive as that carried out for the mainland UK. As a result, various and increasingly complex systems for establishing inventories have been developed which often rely at some level on numerically 'grading' identified sites (as discussed by Brilha 2016).

At one level such approaches can be very helpful as they can enable some form of site identification and classification by a non-specialist—including in areas which, unlike the UK, do not have a wealth of scientific publications. However, at the same time, a lack of specialist scientific understanding by any assessor can have serious consequences—for instance, if a unique site with great scientific *potential* is discarded, as it did not score highly enough against any multidisciplinary (e.g. additional tectonic, mineralogical, etc., features) or touristic value.

(a) (b) (c)

Fig. 2.40 **a** Basaltic rocks used to cover the exterior walls of a restaurant building in Puerto Villamil. *Photo* D. F. Kelley **b** Flat slabs of pahoehoe type lava flow surfaces have been collected and used to tile this low wall in a public area in Puerto Villamil. *Photo* D. F. Kelley **c** flat slabs of pahoehoe type lava flow surfaces have been collected and used to tile this wall on a hostal in Puerto Villamil. *Photo* D. F. Kelley

Table 3 in Brilha (2016) for assessing 'Scientific Value' perfectly demonstrates this potential problem, where, for instance, high scores of "4" are awarded to sites with GSSP or ASSP status (i.e. international stratotypes; "*Criteria B. Key Locality*"), sites with "*more than 3 types of distinct geological features*" ("Criteria E. Geological diversity") and under 'Criteria G' (="use limitations"). Under such a process, however, the globally famous and iconic Middle Cambrian Burgess Shale fauna of British Columbia,Canada—a World Heritage site listed by UNESCO (http://burgess-shale.rom.on.ca/en/; https://www.burgess-shale.bc.ca/) would score only 2 points under Criteria B as it is used by international science but is NOT a GSSP etc., only 1 point under Criteria G, as "*Sampling* [is] *is very hard to be accomplished due to…legal permissions* [e.g. a very high level of protection in a National Park], *physical barriers..*" [e.g. it is half way up a mountain] and perhaps would not score at all under Criteria E, as only one, not two, distinct geological features are present [e.g. a Middle Cambrian fauna only]!

Clearly, some level of comprehensive, expert scientific consultation on any provisional listing of sites for conservation should pick up such issues, but this may be beyond the possibilities of the inventory project in place.

In the case of the Galapagos, the most conspicuous geological heritage is the volcanic heritage of the islands, which represents both past (e.g. 'fossil') and active systems. These are the features and systems which will be considered in greatest detail here, aided by the considerable advances in recent years in studies of Volcanic Geoheritage globally, including as promoted by IAVCEI (the International Association of Volcanologists and Chemistry of the Earth's Interior; https://www.iavceivolcano.org/) and illustrated by monographic works by Erfurt-Cooper (2014), Moufti and Nemeth (2016), etc. and the special volume of *Geoheritage* edited by Németh et al. (2017).

Other categories of Geoheritage importance are also of course present across the archipelago, including coastal erosional and depositional systems, lagoonal systems and sediments, springs, soils and 'cave' deposits in lava tunnels. As little systematic work appears to have been dedicated to these other categories of geological features and systems, an attempt at a comprehensive account would be premature, but should certainly be considered a priority for future study and inventory.

2.3.3 Towards an Inventory of the Geoheritage of the Galapagos

Although no systematic survey of geodiversity features across the Galapagos islands has been carried out, the listing below (as also discussed in Sect. 2.2.5) best represents the general categories of features associated with Galapogean volcanic landscapes. With regard to large-scale features such as shield volcanoes or calderas, the list can be considered complete. However, it should be noted that with the smaller scale features that are quite widespread in a volcanic setting, such as lava flows, tumuli, hornitos, fumeroles, etc., that this list is biased toward those localities that are approved GNP sites and thus are quite accessible. The volcanoes of the western islands in particular are both the most active and the least accessible—both factors that mean there are likely a very high number of some of these smaller scale features, and it may be impractical to document them all individually, although comprehensive aerial survey would still be possible.

As the majority of these sites lie within the Galapagos National Park, however, aspects of their conservation and management will form part of the more general management regimes in place (see Chap. 4). However, as some lie out of the Park, for instance close to human settlement in the case of most quarries, some form of overall geodiversity assessment is still necessary, including to help avoid inadvertent damage—something which could still occur within areas of the Park, however, as an unforeseen bi-product of management activities for ecological features or visitor facilities. There is also the question of what really are the key geological sites of ongoing scientific importance across the islands, for instance those of the highest geoscientific importance, historical importance (e.g. in connection with Darwin's visit), or with potential for current and future research. The identification of such sites can also help inform National Park Management policies, including in relation to facilitating access for continued study and sampling.

As indicated above, various methodologies exist to aid the establishment of inventories of geological features—or geodiversity audits, a term often used in the UK. On one side such audits can attempt to objectively inform on the scientific 'value' of an area, site or feature, on the other they can help assess suitability for educational or touristic use. Either, way, however, conservation and management priorities are informed. Brilha (2016) provides a useful discussion of the inventory process for assessing geological sites, including as a 'Geoheritage' for conservation. However, as indicated previously, the emphasis on a numerical grading can become counterproductive if not adequately scientifically informed, leading to an emphasis on 'geo-diverse' and accessible sites rather than those which really are of international scientific importance for specific features or processes.

Reynard's et al's review for the assessment of 'geomorphosites', however, is also relevant as an inventory of a volcanically active region is as much about landscapes as it is rock types and geological time scales. Not surprisingly, therefore, it was adopted by Tefogoum et al. (2014) in their assessment of volcanoes associated with the Cameroon line in Cameroon.

As in many previous studies, both Brilha (2016) and Reynard et al. (2017) use broadly similar categories for site assessment and hence selection for conservation activity, however, some blurring of objective scientific qualities with management conditions and perceptions of diversity still pervades. These latter qualities might be crucial for the effective management of a site for future use, and its suitability for 'promotion', both educational and touristic, but certainly have no place in any attempt at establishing a primary scientific importance (as opposed to a potential for future scientific study), i.e. only complete loss of the key features would be relevant, as simple concealment, even under urban development, does not mean that the site no longer possesses a scientific importance, or potential in the future! Crucially, in the context of areas such as the Galapagos National Park, which are largely free of any significant human degradation, purely scientific criteria can be meaningfully applied as management regimes can be assumed to be neutral. Outside of the designated park, however, such considerations may become more significant.

Nevertheless, only four basic criteria are necessary for an assessment of 'Scientific Value' and hence 'grading' of, for instance, volcanogenic features to inform ongoing management priorities both within and outside of the GNP, and these are listed/suggested below. Crucially, it is essential that the development of a credible scheme for grading geodiversity sites across the islands will require expert input from active researchers on both the area itself, and volcanic island systems elsewhere in similar tectonic settings.

1. Rarity or representativeness of the features present, with respect to the scientific framework under consideration (e.g. volcanic geodiversity, coastal processes, etc) and geographical study area, ranging from common and frequent (i.e. typical or representative of the framework), to rare and unique. Although the latter categories will often be considered to be of the highest relative scientific value, within any scientific framework, the most representative sites can sometimes be of the greatest significance for ongoing research. Any definition of what might be considered common, including representative, or rare in the context of the Galapagos islands will require expert scientific advice, including in comparison with other similar regions globally (see theme 2, below).
2. Significance/level of scientific interest (actual or potential, for instance as demonstrated by peer-reviewed scientific literature), e.g. local (e.g. by island), regional (e.g. within the Galapagos province), National (within Ecuador), International (e.g. in comparison with adjacent countries in South and Central America, Global (e.g. across different continents, hence of highest scientific value).
3. Integrity and completeness—The state of preservation or overall condition of the scientific features present, can often be of great significance for its scientific value. For instance, an intact and complete spatter cone would normally be considered of higher scientific value that one degraded or eroded. However, both intact and well preserved cinder cones, as well as those eroded to show internal structures, could both be of equivalent scientific value or even the latter superseding the former. Again expert geoscientific opinion is needed.
4. Presence of regional, national, international or global reference features. Such sites are fundamental to the scientific method, and include mineral type localities, national and global stratigraphical reference sections and 'classic' localities, fundamental to the history or the geological sciences.

Other considerations, such as 'Geological diversity' and 'Use limitations' (e.g. access restrictions) which are often invoked by various authors, should not influence any primary scientific assessment as they are more closely related to considerations of educational or touristic potential. For instance, even the case of the highly protected Canadian Burgess Shale lagestatten mentioned previously, for genuine scientific study, there are conservation and legal mechanisms in place through which applications for permission for scientific studies can still be made. Crucially, however, if 'Use limitations' were invoked within the GNP as part of an assessment of scientific value, the great majority of sites present would fail this criteria, due to the high level of restrictions in place!

Similarly, 'additional values' (sensu Reynard et al. 2007, 2016) such as 'Ecological Value', 'Aesthetic Value' and 'Cultural Value' should not be allowed to modify any attempt at an objective geoscientific assessment, although should of course be considered as any future management regime is developed.

Stages in a comprehensive inventory process for the Galapagos should include:

1. Comprehensive topographic survey using stereo-imaged aerial photographs and/or digital elevation modelling. Note that the vertical resolution of readily available sources such as Google Earth (https://earth.google.com/web/) with vertical intervals in several 10 s of meters, or more, will not be sufficient to resolve smaller features such as hornitos, although Lidar imaging which can have a resolution of 20 cm would be ideal (https://en.wikipedia.org/wiki/Lidar; www.lidar-uk.com). Some development of a classification of features themes would be necessary at this stage, for instance as outlined in Sect. 2.2, but with reference to surveys of similar regions elsewhere, but the following could act as a provisional listing or characterisation of features:
   (a) Shield Volcanoes
   (b) Calderas
   (c) Cinder Cones
   (d) Tuff Cones
   (e) Lava Flows (classified by age, type and morphologies, e.g. pahoehoe, aa, pillow lavas, mineralogical type, hornitos, squeeze-up structures, etc)
   (f) Lava Tunnels
   (g) Mineralogy/petrology (e.g. identification of unusual mineralogy/mantle xenoliths, etc) (including moveable geodiversity, i.e. ex situ in institutions)
   (h) Kipukas
   (e) Geothermal springs and fumaroles
   (i) Eruption history/record of phases
   (j) Soils
   (k) Non-volcanic landforms and fluvial geomorphology
   (l) Coastal geomorphology (including records of sea-level change)
   (m) Palaeontological heritage (including moveable geodiversity, i.e. ex situ in institutions)
   (n) Geoheritage in a cultural context (e.g. building materials).
2. Correlation of topographic survey with scientific descriptions and literature (both published and unpublished)
3. Identification of key features for '*ground truthing*' and further survey
4. Development of criteria for the classification and grading of identified features from internationally important to representative.
5. Development of management principles and plans, including for specific sites or areas of particular significance and the establishment of general principles for the

management of different types of feature (hornitos, lava tunnels, kipukas, etc.).

6. Development of a framework for designation or classification of graded sites, including integration within GNP schemes, such as approved sites for general visitors.

In conclusion, the systematic compilation of an inventory of the Galapagos geological features is long overdue, and should be considered to be priority for informing future management of both the GNP and adjacent non-GNP areas. Part of this process requires the development of scheme through which to 'grade' the intrinsic scientific value of the sites and features identified. Only expert advice from relevant specialists can make this scheme credible, even if the actual survey is carried out by others.

2.3.4 Educational Geosites in the Galapagos

Assessing educational and touristic value is a much more complex matter, however, especially within the National Park, as management and access considerations become very significant and typically prohibitive. Brilha (2016) and Reynard et al. (2017), like many other authors, provide schemes through which such potential can be assessed, such as 'Accessibility', 'Security', 'Tourism infrastructure', 'Interpretative facilities', etc. Reynard et al, however, crucially state that: "*…we consider that the potential for educative activities or geotourism as well as the needs for protection are not part of the "quality" of the site and are not, therefore, to be considered as a value*."

The sites that can be visited within the GNP are listed and controlled. Any visit by those ranging from casual tourists, to commercial tour companies, to educational groups from universities, must build their itinerary within the parameters of approved GNP sites, which is also likely to require the presence of a registered guide. While there are many approved visitor sites in the islands, only a subset of them might be considered of geological heritage significance, by demonstrating the geodiversity listed above. All such sites on each island and suggested itineraries are detailed in Chap. 5. Significantly, however, in the absence of a systematic inventory of geological sites across the Galapagos, the only site listing or categorization that does exist is the GNP list. Their accessibility is their primary consideration for selection, a category which Reynard et al. would not consider to be a scientific value.

Most visitors to the Galapagos Islands are there because of the unique scientific characteristics and the importance that Charles Darwin's visit played in the development of our understanding of natural science. This is a true locality for 'geotourism' or 'ecotourism'. Aside from, or perhaps complementary to, the intent of most visitors to have an informative visit to the Galapagos Islands, the aim of the regulations of the GNP for guiding tourists is also that visitors are educated about the natural science of this place. All approved sites must be visited accompanied by a licensed GNP guide. The guides are there to ensure the safety of visitors and the protection of the geological and ecological aspects of the sites, but also to educate the visitors. Education is built into all tours, and in this way, the continued protection of the natural value islands is reinforced.

Most tourists who visit the island have goals of seeing giant tortoises, blue-footed boobies, marine iguanas, sea turtles and finches; or perhaps to experience the marine ecosystem with snorkeling and scuba diving tours. The volcanic geology of the islands is not particularly well known or targeted by visitors. However, as described in Kelley and Salazar (2016) the Galapagos Islands are an ideal place to conduct a field-based geology course. Generally, a well-designed geology education tour of the islands will include visits sites on San Cristobal, Santa Cruz, and Isabela Islands with the goal of providing participants with a sense of the plate tectonic control on the distribution of ages of volcanic activity through the island chain. Also, an itinerary should include a list of sites that allow for viewing of as much of the volcanic geodiversity that the islands offer as possible within the time frame of the visit. In Chap. 5 we review the subset of sites of most geological value and describe location, accessibility, and logistical considerations of each (Fig. 2.41).

2.4 Designating Geological Sites for Conservation and Public Use

2.4.1 International Designations

Introduction

Every national conservation system has its structures and categories of designations, typically reflecting a spectrum from the most important sites with the highest levels of protection to areas with much less restriction. As around 97% of the land area of the Galapagos is protected within the Galapagos National Park, the islands have one of the highest levels of protection of anywhere in the world, and not just within Ecuador. As is typical of such areas, specific features, especially geological do not have a separate status or even listing.

Beyond national conservation systems, with varying degrees of effectiveness, especially outside of highly protected national park areas, lie a range of international designations of global applicability, although all relying on national systems to operate. Most widely known of which is UNESCO's (United Nations Educational Scientific and Cultural Organization) recognition of sites of 'World Heritage' status', a category which includes both cultural and

Fig. 2.41 Educational use of a Galapagos geosite; Sierra Negra caldera rim. *Photo* D. F. Kelley

natural sites. In addition, a new UNESCO program, Global Geoparks, is explicitly geologically focused, and aims to promote the sustainable use of geologically important areas. Purely scientific designations for key geological sites—or rather 'listings'—are also developing, promoted by international scientific organizations such as IUCN (International Union for the Conservation of Nature) and IUGS (International Union of Geological Sciences).

UNESCO listed World Heritage sites

The 'ultimate' global confirmation of a site or region's importance is of course listing by UNESCO as ***World Heritage***. The designation originated in UNESCO's 1972 *Convention for the Protection of the World Cultural and Natural Heritage* which states:

> ***Noting*** *that the cultural heritage and the natural heritage are increasingly threatened with destruction not only by the traditional causes of decay, but also by changing social and economic conditions which aggravate the situation with even more formidable phenomena of damage or destruction,*
>
> ***Considering*** *that deterioration or disappearance of any item of the cultural or natural heritage constitutes a harmful impoverishment of the heritage of all the nations of the world,*
>
> ***Considering*** *that protection of this heritage at the national level often remains incomplete because of the scale of the resources which it requires and of the insufficient economic, scientific, and technological resources of the country where the property to be protected is situated,*
>
> ***Recalling*** *that the Constitution of the Organization provides that it will maintain, increase, and diffuse knowledge, by assuring the conservation and protection of the world's heritage, and recommending to the nations concerned the necessary international conventions,*
>
> ***Considering*** *that the existing international conventions, recommendations and resolutions concerning cultural and natural property demonstrate the importance, for all the peoples of the world, of safeguarding this unique and irreplaceable property, to whatever people it may belong,*
>
> ***Considering*** *that parts of the cultural or natural heritage are of outstanding interest and therefore need to be preserved as part of the world heritage of mankind as a whole,*
>
> ***Considering*** *that, in view of the magnitude and gravity of the new dangers threatening them, it is incumbent on the international community as a whole to participate in the protection of the cultural and natural heritage of outstanding universal value, by the granting of collective assistance which, although not taking the place of action by the State concerned, will serve as an efficient complement thereto,*
>
> ***Considering*** *that it is essential for this purpose to adopt new provisions in the form of a convention establishing an effective system of collective protection of the cultural and natural heritage of outstanding universal value, organized on a permanent basis and in accordance with modern scientific methods,*
>
> ***Having decided****, at its sixteenth session, that this question should be made the subject of an international convention,*
>
> ***Adopts*** *this sixteenth day of November 1972 this Convention.*

Article 2 defines the natural aspect of heritage in the context of the Convention

> *For the purposes of this Convention, the following shall be considered as "natural heritage:*

> ***natural features*** *consisting of physical and biological formations or groups of such formations, which are of outstanding universal value from the aesthetic or scientific point of view;*
> ***geological and physiographical formations*** *and precisely delineated areas which constitute the habitat of threatened species of animals and plants of outstanding universal value from the point of view of science or conservation;*
> ***natural sites*** *or precisely delineated natural areas of outstanding universal value from the point of view of science, conservation or natural beauty."*

Crucially, "*Geological and physiographical* [i.e. geomorphological] *formations", "natural features consisting of physical... formations or groups of such formations*" and other "*natural sites*" of "*outstanding universal value"* from a scientific, conservation or aesthetic perspective are all explicitly included. By 2005 these categories had evolved into the following 10 Selection Criteria (https://whc.unesco.org/archive/opguide05-en.pdf) which remain the framework for all World Heritage nominations and listings (https://whc.unesco.org/en/criteria/). The proposed site, feature or artefact should, therefore, be selected:

(i) "*to represent a masterpiece of human creative genius;*"
(ii) "*to exhibit an important interchange of human values, over a span of time or within a cultural area of the world, on developments in architecture or technology, monumental arts, town-planning or landscape design;*"
(iii) "*to bear a unique or at least exceptional testimony to a cultural tradition or to a civilization which is living or which has disappeared;*"
(iv) "*to be an outstanding example of a type of building, architectural or technological ensemble or landscape which illustrates (a) significant stage(s) in human history;*"
(v) "*to be, an outstanding example of a traditional human settlement, land-use, or sea-use which is representative of a culture (or cultures), or human interaction with the environment especially when it has become vulnerable under the impact of irreversible change;*"
(vi) "*to be directly or tangibly associated with events or living traditions, with ideas, or with beliefs, with artistic and literary works of outstanding universal significance. (The Committee considers that this criterion should preferably be used in conjunction with other criteria);*"
(vii) "*to contain superlative natural phenomena or areas of exceptional natural beauty and aesthetic importance;*"
(viii) "*to be outstanding examples representing major stages of earth's history, including the record of life, significant on-going geological processes in the development of landforms, or significant geomorphic or physiographic features;*"
(ix) "*to be outstanding examples representing significant on-going ecological and biological processes in the evolution and development of terrestrial, fresh water, coastal and marine ecosystems and communities of plants and animals;*"
(x) "*to contain the most important and significant natural habitats for in situ conservation of biological diversity, including those containing threatened species of outstanding universal value from the point of view of science or conservation.*"

The listing of the Galapagos as a World Heritage site in 1978 is a testament to its iconic contribution to global studies and understanding of evolutionary processes, as first discussed by Charles Darwin. The criteria for approval of World Heritage status were its "outstanding universal value", including Darwin's findings, the number of species, the large proportion of endemics, and significant concentrations of plants and animals which are rare or endangered, and its "integrity" (most notably that such a large percentage of the area of the islands was protected as the GNP. The justification for listing under each criteria is explained by UNESCO as follows (taken from https://whc.unesco.org/en/list/1):

"**Criterion vii**: *The Galapagos Marine Reserve is an underwater wildlife spectacle with abundant life ranging from corals to sharks to penguins to marine mammals. No other site in the world can offer the experience of diving with such a diversity of marine life forms that are so familiar with human beings, that they accompany divers. The diversity of underwater geomorphological forms is an added value to the site producing a unique display, which cannot be found anywhere else in the world.*"

"**Criterion viii**: *The archipelago's geology begins at the sea floor and emerges above sea level where biological processes continue. Three major tectonic plates—Nazca, Cocos and Pacific—meet at the basis of the ocean, which is of significant geological interest. In comparison with most oceanic archipelagos, the Galapagos are very young with the largest and youngest islands, Isabela and Fernandina, with less than one million years of existence, and the oldest islands, Española and San Cristóbal, somewhere between three to five million years. The site demonstrates the evolution of the younger volcanic areas in the west and the older islands in the east. On-going geological and geomorphological processes, including recent volcanic eruptions, small seismic movements, and erosion provide key insights to the puzzle of the origin of the Galapagos Islands. Almost no other site in the world offers protection of such a complete continuum of geological and geomorphological features.*"

"**Criterion ix**: *The origin of the flora and fauna of the Galapagos has been of great interest to people ever since the publication of the "Voyage of the Beagle" by Charles Darwin in 1839. The islands constitute an almost unique* *example of how ecological, evolutionary and biogeographic processes influence the flora and fauna on both specific islands as well as the entire archipelago. Darwin's finches, mockingbirds, land snails, giant tortoises and a number of plant and insect groups represent some of the best examples of adaptive radiation which still continues today. Likewise, the Marine Reserve, situated at the confluence of 3 major eastern Pacific currents and influenced by climatic phenomena such as El Niño, has had major evolutionary consequences and provides important clues about species evolution under changing conditions. The direct dependence on the sea for much of the island's wildlife (e.g. seabirds, marine iguanas, sea lions) is abundantly evident and provides an inseparable link between the terrestrial and marine worlds."*

"**Criterion x:** *The islands have relatively high species diversity for such young oceanic islands, and contain emblematic taxa such as giant tortoises and land iguanas, the most northerly species of penguin in the world, flightless cormorants as well as the historically important Darwin's finches and Galapagos mockingbirds. Endemic flora such as the giant daisy trees Scalesia spp. and many other genera have also radiated on the islands, part of a native flora including about 500 vascular plant species of which about 180 are endemic. Examples of endemic and threatened species include 12 native terrestrial mammal species (11 endemic, with 10 threatened or extinct) and 36 reptile species (all endemic and most considered threatened or extinct), including the only marine iguana in the world. Likewise the marine fauna has an unusually high level of diversity and endemism, with 2,909 marine species identified with 18.2% endemism. High profile marine species include sharks, whale sharks, rays and cetaceans. The interactions between the marine and terrestrial biotas (e.g. sea lions, marine and terrestrial iguanas, and seabirds) are also exceptional. Recent exploration of deep sea communities continues to produce new additions to science.*"

UNESCO Global Geoparks

The concept of using geological heritage as a basis for sustainable economic and social development was perhaps first fully developed within the Réserve Geologique de Haute-Provence, in SE France (www.resgeolo4.org) from the late 1980s. Here a combination of dramatic scenery and a range of primarily palaeonotologically-important protected sites was built into a sophisticated presentation of geological heritage, which began to attract thousands of visitors each year and local communities benefited through providing accommodation, restaurants and other tourist facilities.

The development of collaborative links with three other protected geological areas across Europe (Lesvos Petrified Forest in Greece (www.petriefiedforest.gr), the Geopark Gerolstein/Vulkaneifel in Germany (http://www.geopark-vulkaneifel.de) and Parque Cultural de Maestrazgo in Spain (www.maestrazgo.org), and UNESCO's then Earth Science Branch, established the *European Geoparks Network* in 2000, all using European Community LEADER II funding to support the sustainable development of disadvantaged areas.

By September 2007, 32 European Geoparks in 13 countries had been admitted to the Network (www.europeangeoparks.org) and eventually, despite initial reluctance, UNESCO formally recognised Geoparks as a global designation with the following definition (www.unesco.org/science/earthsciences/geological_heritage):

- *"Geoparks are defined as sites or areas of geological significance, rarity or beauty, in which geological features play a significant part, and where the geological heritage is protected and developed at the same time".*
- *"Geoparks shall foster socio-economic development that is culturally and environmentally sustainable by triggering tourism in the form of eco- or geo-tourism. Like this, a Geopark can have a considerable potential for economic development in rural and less developed regions. The management body of a Geopark shall take care of the logistic support for environmental education and training, research and monitoring, related to issues of conservation and sustainable development."*

In 2016, however, as the movement had developed from strength to strength, UNESCO formally adopted Global Geoparks as a programme, celebrating the new designation, the first since the World Heritage Convention of 1972, at a Global congress held in the English Riviera Global Geopark (www.englishrivierageopark.org.uk) in Torbay (Devon), South West England (http://www.unesco.org/new/en/natural-sciences/environment/earth-sciences/unesco-global-geoparks/).

With the formation of the Latin America and Cribbean Geoparks Network (GEOLAC) which held its first meeting in the Comarca Minera Hidalgo UNESCO Global Geopark in Mexico in January 2018 (http://globalgeoparksnetwork.org/?p=2416) there is more support for the initiative throughout the region. Crucially, October 2018 saw the 2nd Ecuadorian Geoparks meeting take place in Tena, Napo Province (http://www.unesco.org/new/fileadmin/MULTIMEDIA/Quito/pdf/Primera_circular.pdf) to cleberate the three Ecuadorain Geopark projects then active (Imbabura Geopark (https://www.andes.info.ec/es/noticias/news/1/unesco-ecuador-geopark), Volcano Tungurahua Geopark (https://www.researchgate.net/project/UNESCO-GGN-Dossier-for-the-Aspiring-Geopark-Volcano-Tungurahua-Ecuador) and Geoparque Napo Sumaco (https://www.facebook.com/NapoSumaco)) (Fig. 2.42).

Fig. 2.42 Imbabura and Volcano Tungurahua—Geopark projects in Ecuador **a** Tungurahua Volcano *Photo* D. F. Kelley; **b** Imbabura Volcano *Photo* D. F. Kelley: **c** Map of these two localities to the north and south of Quito, Ecuador

International Union for the Conservation of Nature (IUCN): motions and resolutions:

The IUCN is the most influential global organization in the field of nature conservation, and hence, through the adoption of international motions and recommendations, can strongly influence the development of international and national scientifically-informed policy.

A major milestone in the development of geological conservation as a mature subject within global natural nature conservation effort was, therefore, the adoption of a formal resolution at the IUCN World Conservation Congress, held in Barcelona (Cataluña, Spain) in October 2008. The preamble to *WCC 2008 Resolution4.040* on the *Conservation of geodiversity and geological heritage* (https://portals.iucn.org/library/sites/library/files/resrecfiles/WCC_2008_RES_40_EN.pdf) states:

> *'The main objective of this resolution is to incorporate the conservation of geodiversity and geological heritage into the agenda of IUCN, the main reason being that both are part of the Earth's natural heritage and hence need to be considered by IUCN. In order to achieve this, the preamble draws attention to (1) the conceptual framework set by the World Heritage Convention in 1982, which considers geological heritage as part of the natural heritage, (2) the objectives of the current International Year of Planet Earth adopted by the UN General Assembly, and (3) the pioneering steps set by the recommendations of the European Council in 2004."*

The preamble also recalls basic concepts regarding geodiversity and geological heritage, and the need to consider these aspects in nature conservation, land management and sustainable use of resources:

> *"It is often forgotten that all we know about Earth's evolution, including the evolution of climate, species, habitats and resources, is based on the geological record. Geological heritage includes those most valuable sites with the best record of Earth's evolution."*
>
> *"Likewise, the diversity of geological and geomorphological features underpins biological, cultural and landscape diversity, and thus needs to be considered as one more value of natural heritage requiring appropriate adaptive management towards an integrated conservation."*
>
> *"We believe the time is appropriate for IUCN to begin considering geodiversity and geological heritage in forums and congresses. Our focus with this resolution at this fourth World Conservation Congress is to promote actions and initiatives in this direction, with the hope that future work may gradually develop towards recommendations regarding the conservation of geodiversity and geological heritage. The general objective of this motion strictly follows the general objective of IUCN: to promote the conservation and sustainable use of natural heritage for future generations. But, in order to be complete, natural heritage needs to include geological heritage, the natural archive for the memory of the Earth."*

The Resolution itself states as follows:

> *"NOTING that the United Nations General Assembly proclaimed 2008 to be the International Year of Planet Earth, initiated jointly by the International Union of Geological Sciences (IUGS) and the United Nations Educational, Scientific and Cultural Organization (UNESCO) in order to increase awareness of the importance of Earth sciences in achieving sustainable development and promoting local, national, regional and international action;*
>
> *AWARE of the rapidly growing interest and commitment of States, NGOs, and communities to save, study and sustainably use their geodiversity and geological heritage;*
>
> *RECALLING that geodiversity, understood to include geological and geomorphological diversity, is an important natural factor underpinning biological, cultural and landscape diversity, as well as an important parameter to be considered in the assessment and management of natural areas;*
>
> *RECALLING FURTHER that geological heritage constitutes a natural heritage of scientific, cultural, aesthetic, landscape, economic and/or intrinsic values, which needs to be preserved and handed down to future generations;*
>
> *NOTING the pioneering experience led by the United Nations Educational, Scientific and Cultural Organization (UNESCO) and other international institutions in promoting the conservation and sustainable use of geological heritage through the development of the Global Geoparks Network (GGN);*
>
> *RECOGNIZING the escalating impact of development that is frequently unsustainable upon the world's geodiversity and geological heritage;*
>
> *RECOGNIZING FURTHER that in planning such development, the intrinsic values, both material and intangible, of the geodiversity, geoheritage and geological processes present at natural areas are often underestimated or even ignored;*
>
> *AWARE that the Global Geopark Network and Global Geosites Program of UNESCO cover less than 1% of the world's land surface and less than 1% of the marine area, and that most of the geological heritage lies in the wider landscape outside protected areas;*
>
> *RECALLING that the Preamble to the World Heritage Convention adopted by the UNESCO General Conference recognizes that the deterioration or disappearance of any item of the natural heritage constitutes a harmful impoverishment of the heritage of all the nations of the world, and that Article 2 considers geological and physiographical formations of outstanding universal value from the point of view of science or conservation as natural heritage;*
>
> *RECALLING ALSO the pioneering trend set by the adoption of Rec(2004)3 (Conservation of the Geological Heritage and Areas of Special Geological Interest) by the Council of Europe in 2004, and its call to strengthen cooperation amongst international organizations, scientific institutions and NGOs in the field of geological heritage conservation, and participate in geological conservation programmes;*
>
> *RECOGNIZING that the conservation of geodiversity and geological heritage contributes to deal with species loss and ecosystem integrity;*
>
> *NOTING that the IUCN guidelines for applying protected area management categories explicitly consider amongst the objectives common to all protected areas the need to: (a) maintain diversity of landscape or habitat, (b) conserve significant landscape features, geomorphology and geology, and (c) conserve natural and scenic areas of national and international significance for cultural, spiritual and scientific purposes;*
>
> *RECALLING that the conservation of geodiversity and geological heritage at international, national and local levels contributes to the objectives of the United Nations Decade of Education for Sustainable Development (2005–2014);*
>
> *RECOGNIZING the important role of geological and geomorphological conservation in maintaining the character of many landscapes;*

RECOGNIZING ALSO that the conservation and management of geological heritage need to be integrated by governments in their national goals and programmes;

NOTING that some areas with geological and geomorphological values will deteriorate if they are not taken into account in planning and development policies; and

AWARE of the need to promote the conservation and appropriate management of the world's geological heritage, in particular areas of special geological interest;

REQUESTS the Director General to: (a) Convene a continuing series of meetings on Geodiversity and Geological Heritage in the regions in partnership with members and other organizations; and (b) Establish a Secretariat focal point to facilitate the organization of these meetings and to provide their continuity while maintaining the minimum organization and administration possible; and

CALLS ON IUCN's Commissions, especially the World Commission on Protected Areas, to support the Secretariat in the design, organization, hosting and funding of future Forum sessions on Geodiversity and Geological Heritage to ensure that this mechanism will achieve the widest possible involvement of government, independent sector groups, and international organizations around the world."

The origins of this motion lie in the Council of Europe's 2003 '*Recommendation Rec (2004)3 on Conservation of the Geological Heritage and areas of Special Geological Interest*' (http://jncc.defra.gov.uk/pdf/councilofeurope1.pdf) which in turn can be traced back to the original '*Digne declaration*' on the '*Rights of the Memory of the Earth*' of 1991 (http://www.progeo.ngo/downloads/DIGNE_DECLARATION.pdf) which represents the first significant international statement to raise awareness of the importance to our societies of geological and geomorphological heritage. (see Chap. 4 also).

Following the 2008 resolution, a Geological Specialist Group (GSG) was established within IUCN's World Commission for Protected Areas (WCPA) (https://www.iucn.org/theme/protected-oareas/wcpa/what-we-do/geoheritage) to provide specialist advice on all aspects of geodiversity to IUCN in relation to protected areas (including World Heritage sites) and their management. The GSG subsequently ensured that Geodiversity and Geological Heritage was incorporated into IUCN's formal work programme in 2012 through resolution WCC-2012-Res-048-*Valuing and conserving geoheritage within the IUCN Programme 2013–2016* (https://portals.iucn.org/library/sites/library/files/resrecfiles/WCC_2012_RES_48_EN.pdf). Another major achievement was the adoption at the World Conservation Congress, held in Hawaii in September 2016, of resolution WCC-2016-Res-083-EN on the *Conservation of moveable geological heritage*—one of the most difficult aspects of geological heritage to manage, as fossil and mineral specimens often leave national legal jurisdictions as part of an uncontrolled international trade.

From 2015, the GSG has developed a chapter on geoconservation for IUCN's "*Protected Area Governance and Management*" handbook for (Worboys et al. 2015; available at: http://press-files.anu.edu.au/downloads/press/p312491/pdf/book.pdf?referer=372) and continues to develop best practice in this area. In addition, the GSG is currently (2018) completing a review of volcanic geoheritage within World Heritage sites, which will be of particular significance to the Galapagos.

Crucially, from 2016, IUCN has developed a concept of Key Biodiversity Areas (KBAs) as "*sites contributing significantly to the global persistence of biodiversity', in terrestrial, freshwater and marine ecosystems.*" (https://portals.iucn.org/library/node/46259; http://www.keybiodiversityareas.org/home) and from 2018 is seeking to develop a parallel suite of "*Key Geodiversity Areas*", to represent the world's most important Geoheritage sites (Woo et al. 2018).

International Union of Geological Sciences (IUGS) projects, programs and the International Commission for Geoheritage:

The IUGS is one of the of the largest and most active NGO scientific organizations in the world and promotes and encourages the study of geological problems of direct interest to society, governments, industry, and academic groups by supporting and facilitating international and interdisciplinary cooperation. Key areas of activities include developing international standards, Geoscience education, Geoscience information provision and advice on Environmental management and hazards.

IUGS has had a long involvement with aspects of Geodiversity and Geoheritage, going back to the development of the concept of Global Geosites in the 1990s. The project developed from the initial work of the Working Group on Geological and Palaeobiological sites, a collaboration between IUGS, UNESCO, IUCN and the IGCP (the then International Geological Correlation Programme which was affiliated with IUGS). The aim of the project was to produce a 'Global Indicative List of Geological Sites' (or 'GILGES') to inform World Heritage site selection (Cowie 1993). By 1994 this project had developed into a concept of 'Global Geosites' a much broader initiative, beyond simply identifying potential World Heritage sites.

The project aimed to involve the global geological community in providing inventory and data to support national and global outreach to protect geological resources for scientific and education objectives. These aims were designed to mesh with national and global initiatives and address the issue of how best to represent the diversity and richness of key geoscientific sites. By 1995, IUGS, with the support of UNESCO, promoted a project to compile a global inventory and database of key Geoheritage sites through the formation of a Global Geosites Working Group (GGWG) with the following terms of reference (Wimbledon 1996, Wimbledon et al. 1999, 2000):

(1) To compile the global Geosite inventory base on the scientific assessment of key geo(morphological)sites,
(2) To compile the Geosites IUGS database of key sites and terrains,
(3) To use the Geosites inventory to further the cause of geological conservation and support geological science in all its forms,
(4) To support regional or national initiatives aiming to compile comparative inventories,
(5) To participate in and support meetings and workshops that examine site selection criteria, selection methods or conservation of key sites,
(6) To assess the scientific merits of sites in collaboration with specialists, research groups, associations, commissions, subcommissions etc.
(7) To advise IUGS and UNESCO on the priorities for conservation in the global context, including World Heritage candidate sites.

To aid this process a range of Criteria were also developed to aid site selection, including to determine international scientific importance. Although the IUGS project global stumbled, its principles were widely adopted and applied international, especially in Europe through the work of ProGEO (The European Association for the Conservation of the Geological Heritage; http://www.progeo.ngo/) and perhaps most notably in Spain (Garcia-Cortés et al. 2009; available at http://www.igme.es/patrimonio/GEOSITES/0_INDEX.pdf).

Subsequently collaboration with UNESCO, however, led to the establishment of the the International Geoscience and Geoparks Programme, or IGGP, which incorporate a restructured IGCP, now the International Geoscience Programme (www.unesco.org/new/en/natural-sciences/environment/earth-sciences/international-geoscience-programme/) and, crucially, UNESCO Global Geoparks (http://www.unesco.org/new/en/natural-sciences/environment/earth-sciences/unesco-global-geoparks/) as a *joint* venture.

Although UNESCO has taken the highest profile role in Global Geoparks, IUGS maintains an underpinning scientific role as defined in the "*Operational Guidelines for UNESCO Global Geoparks*": (www.unesco.org/new/fileadmin/MULTIMEDIA/HQ/SC/pdf/IGGP_IGCP_UGG_Statutes_Guidelines_EN.pdf):

> *"A UNESCO Global Geopark must contain geology of international significance.""The international significance of the geological heritage of each new UNESCO Global Geopark application will be assessed by desk-top advisors following specific and publicly available scientific criteria.""The UNESCO Secretariat shall liaise with IUGS and other organizations, as appropriate, to obtain independent, desk-top scientific assessments.""IUGS will be asked to coordinate this role and to ensure that all statements on the scientific value and international significance of the geological heritage of an aspiring UNESCO Global Geopark are available annually in time so that evaluators can access them ahead of the field evaluation mission."*

Meanwhile, two Task Groups had become established within IUGS in 2008, one focusing on Geoheritage, the other on Heritage Stones, the latter developing a concept of Global Heritage Stone Resources. The success of the two groups led to their amalgamation, in September 2016 within a new IUGS Commission, the International Commission for Geoheritage (ICG), at the 35th International Geological Congress, held in Cape Town, South Africa, as a Heritage Sites and Collections Subcommission (HSCS; http://geoheritage-iugs.mnhn.fr) and a Heritage Stones Subcommission (HSS; http://globalheritagestone.com/) (Pereira and Page 2016).

Although the development of the full Commission and the HSCS has been slow, it is likely that the latter, in particular, will revive the IUGS concept of 'Global Geosites'. Crucially, however, as IUCN's Geological Specialist Group (GSG) is already beginning to develop 'Key Geodiversity Areas' (Woo et al. 2018)—a concept indistinguishable from 'Global Geosites'—IUGS has already lost the initiative in this area. Nevertheless, the development of a collaboration between the IUCN's GSG and IUGS's, ICG, as well as other key players such as ProGEO, could provide an opportunity to built a scientifically-credible methodology which could be implemented globally through IUCN's influential position in nature conservation. The results of such a collaboration will inform geological conservation initiatives globally and could be of especially significance to the Galapagos.

2.5 Protecting Geoheritage Sites

Within large protected areas such as National Parks, geological features usually benefit from some of the general management guidelines and prescriptions in place, although frequently these may not be specifically derived for such features. The potential remains, therefore, that without some form of inventory process, inadvertent damage can still potentially take place (as indicated previously). Outside the zones of protection of a National Park, typically only additional site-focused conservation systems, which can be legally enforced, will achieve the sensitive management of geological features—although if a local educational or touristic value can be established, individual land-owners and managers can often be persuaded to take on such responsibilities themselves.

Typical threats to geological features in whatever environment they may be found, and management solutions, can be classified as follows (based on NCC 1990, Page 2017):

1. **Natural degradation and vegetation growth**: In tropical and other humid environments, vegetation growth can be a major conservation issue for inland sites, leading to the

obscuring of geological exposures and restricting access to them. Physical damage also commonly results as roots break-up exposures, sometimes causing dangerous instabilities, and rain and frost contribute to the development of screes and talus which can bury the lowest parts of the same exposures. Under certain circumstances the deposits present may also be affected by weathering, as oxygen and water chemically react with some of their component minerals, especially preserved organic materials and sulfides (e.g. pyrite) (Fig. 2.43).

Management solutions: Periodic clearance of vegetation and any accumulated scree is the only practical solution to managing most such sites, should the scientific or educational value of the features present merit it. This can be carried out with hand tools or using mechanical excavators. Where relatively soft and unconsolidated deposits are present, however, serious consideration may need to be given to either burying the exposure temporarily to limit deterioration, or ceasing periodic clearance to allow the exposure to naturally become covered, and therefore similarly protected. In addition, where its extent is limited, regular re-excavation could eventually remove the entire deposit, and under such circumstances is best only carried out for specific research or educational activities—with ongoing conservation activities restricted to maintaining access.

2. **Coastal protection/flood defense**: Eroding coastlines, whilst continuously providing fresh geological materials for geoscientists to study, can provide major problems for coastal zone management, as infrastructure and developed areas such as towns and villages become threatened. An inevitable consequence is the loss of the important geological exposures, as issues of safety and property value outweigh even the strongest scientific and heritage cases for their retention. The issue is particularly extreme where coastlines are relatively 'soft', being composed of poorly lithified deposits, such as Quaternary sediments. Coast protection and flood defence are closely linked issues, although the latter can also apply to inland river systems, where damage to natural processes due to river canalisation,

Fig. 2.43 Densely overgrown lava flow, now inaccessible—including to large tortoises (Galapagos National Park, beside paved path from Pto. Baquerizo Moreno to Bahia Wreck, San Cristobal). *Photo* K. N. Page

floodbanks and dams can lead to not only the loss of specific landforms, but also the fragmentation of the river system as whole. The net result being that an entirely natural system no longer functions (Fig. 2.44).

Management solutions: In coastal locations where infrastructure and development is already at risk, only very, very rarely will any conservation scheme take precedence over economic or political issues. Where structures are not at immediate risk, however, 'soft engineering' options are possible, for instance the replenishment of beach materials or the construction of barriers (i.e. groynes) to restrict its movement. Such works have the advantage of slowing down erosion rates to an 'acceptable' level without permanent concealing or damaging the key geological exposures.

Under certain circumstances, the construction of a sea wall or even promenade at the base of the cliff exposure may be difficult to prevent. Having removed the active marine erosion that previously maintained a natural exposure, however, cliff faces soon degrade as vegetation growth begins. Under such circumstances, conservation management and procedures becomes more akin to that applied to disused artificial exposures, such as quarries and cuttings.

As always, a more strategic approach to managing coastal and river systems is always best, and the control of development in sensitive areas is always preferable to a rearguard conservation campaign, once the infrastructure is already in place.

3. **Waste disposal (landfill and effluent)**: Open excavations can provide convenient opportunities—both legal and illegal—for the disposal of a wide range of bi-products of human activity, ranging from, relatively, inert building materials to potentially contaminating domestic and industrial waste. If the excavation exposes key geological features, such infill will lead to their complete loss. In addition, contaminated ground water can also leach into and seriously damage other adjacent features, such as spring and cave systems.

Management solutions: A rigorously enforced legal framework for regulating waste disposal and dealing with illegal activities is crucial to reducing the risks posed by

Fig. 2.44 Developed and defended coastline (and cinder cone to left); Pto. Baquerizo Moreno, San Cristobal. *Photo* K. N. Page

waste disposal at any site. Where such activity has been authorized, however, the retention of geological exposures may still be possible as a legal condition, for instance using retaining walls or bunds. However, in many cases, documentation of exposures and the recovery of representative specimens for research or education as part of a 'rescue dig' program may be the only conservation solution available.

Occasionally, recreation of the lost exposures may be possible adjacent to the infilled site, for instance by excavating a trench. In the long term, however, only more forward-looking approaches to waste disposal, such as recycling, will reduce the pressure on geological sites as disposal areas.

4. **Mineral/aggregate extraction and restoration of working sites**: Although active quarries create and can maintain important geological exposures, they can also remove or significantly damage key deposits and features, even remove them in their entirety (see Sect. 2.2.5.11). Restoration and landscaping of such sites, on cessation of working—as may be stipulated in certain planning conditions—can also lead to the loss of any exposures created during working, for instance as rock faces are buried or graded or otherwise made inaccessible.

Management solutions: Where working quarries are controlled within spatial planning systems, by far the best way of reducing damage to unique geological features and ensuring the retention of 'conservation sections' (i.e. key exposures) on cessation of working is through the planning system. This may entail the modification of working schemes and restoration plans, together with a range of engineering solutions to ensure rock face stability and accessibility. Where appropriate comprises cannot be reached, however, or the nature of the site makes retention of conservation exposures impractical—for instance where they will ultimately be below ground water level—recording programs and rescue digs can again become important.

Under certain circumstances, however, especially where soft and unconsolidated Quaternary sediments may be present, the only viable conservation option maybe the establishment of a 'Green field' site adjacent to the working site, which could then be accessed through boreholes or temporary excavations.

5. **Civil engineering, industrial and domestic developments and projects**: Development of any form transforms landscapes and has the potential to seriously damage or even obliterate features of geological and geomorphological importance. Damage may be either direct, for instance the removal of a landform, or indirect as a result of the infilling of quarries and natural depressions with waste materials excavated from elsewhere. Such issues are most commonly associated with commercial and domestic building workings and road construction, although other engineering activities such as the construction of airports and dams can have a similar effect, as geological and geomorphological features are removed or concealed.

Management solutions: Informed spatial planning is typically the only way in which the potential damage caused by development to key conservation sites can be avoided—crucially in association with strong enforcement procedures where the activity is technically illegal. Adequate impact assessment procedures are essential, although awareness of geological issues is typically limited amongst both local authority planners and developers. Engineered solutions, which retain rock exposures, can be implemented if such sites can be identified through such procedures. Failing this, or in addition, rescue digs and recording schemes can ensure that at least some of the geological heritage revealed during development can be documented and conserved—at least in a museological sense.

6. **Agricultural and other land management practices, including afforestation**: A range of other management practices have the potential to damage geological and geomorphological sites. Many of these are associated with farming and forestry practices, including the infill or contamination of sites with waste materials, the construction of buildings and access tracks, deep ploughing of delicate surface features and deposits and the development forestry or other plantations, which will obscure landform features.

Changes in management regimes, including for ecological conservation purposes can also become an issue, for instance where grazing levels are reduced and the growth of scrub leads to the obscuring of landform features and exposures, the excavation of ponds for wildlife can damage hydrological and fluvial features and the placing of fencing, gating and other structures and features on rock faces and within caves can lead to further damage (see Fig. 2.34).

Management solutions: More so than with most other categories of site damage, the raising of awareness of geological conservation issues amongst farmers and other land managers is a key issue in avoiding what is so commonly, inadvertent rather than deliberate damage. Better informed agri-environment and ecological conservation schemes are clearly essential, as is improved communications between governmental and non-governmental conservation agencies and trusts. A valuable tool in this process is the development of integrated bio—and geodiversity management and strategic conservation plans, which can be very successfull for the management sites with both biological and geological interests (see Page 2017 and Matthews 2014 for instance).

7. **Overuse or misuse**: In many countries, inappropriate uses of geological sites can continue to be a major issue, even when issues of development control and countryside management regimes have been resolved. Some of this activity is specifically targeted at the geological resource present as fossils and minerals can be highly sort after items, often with a significant international market value. In

extreme cases, the site can be, literally, 'removed'. In other cases, however, the damage may be inadvertent, for instance as a consequence of other recreational activities, such as climbing and caving.

Management solutions: Irresponsible activity by specimen collectors can only ultimately be controlled by the strict enforcement of conservation legislation—with the implication that improved site policing or wardening may be necessary. In many cases, however, improved education of site users and the establishment of Codes of Conduct can be very useful. Wherever site damage through overuse or misuse is a significant issue, some form of signing should be considered essential, to make visitors aware of the problems that they might be causing.

2.6 The Future of Geodiversity Conservation in the Galapagos—Summary

Although the Galapagos is listed by UNESCO under Criteria (vii) to (x) of the Operation Guidelines of the World Heritage Convention, only Criteria (viii) refers specifically to geological features and these are primarily volcanic in origin. However, although a general protection is available for GNP areas, no protective designations appear to apply to any features outside of the Park's boundaries, where by definition, they may be at highest risk form human activities. In addition, as discussed above, without a systematic inventory of geodiversity features across the GNP area, there is no source available against which to prioritize site management activity or inform ecological-focusses management activity (e.g. to avoid inadvertent damage).

Clearly, therefore, a priority should be a mapping of key geodiversity features across the islands, including in both GNP and non-GNP areas, as discussed above. From this database key features of regional, national and international importance can be identified and either designated or delineated to inform future management priorities. Such a database will also help identify new sites for managed access for educational and/or tourism. Crucially, it could also identify keys sites for ongoing, even promoting, scientific activity, and hence enable site managers, in liaison with a broader scientific community, to develop specific access and user guidelines to facilitate future studies.

References

Ali JR, Aitchison JC (2014) Exploring the combined role of eustasy and oceanic island thermal subsidence in shaping biodiversity on the Galapagos. J Biogeographys 41:1227–1241

Asimov PD, Hircschmann MM, Stolper EM (2001) Calculation of peridotite partial melting from thermodynamic models of minerals and melts, IV. Adiabatic decompression and the composition and mean properties of mid-ocean Ridge Basalts. J Petrol 2(5):963–998

Bailey K (1976) Potassium-argon ages from the Galapagos Islands. Science 192:465–467

Barckhausen U, Ranero CR, Von Huene R, Cande SC, Roeser HA (2001) Revised tectonic boundaries in the Cocos plate off Costa Rica: implications for the segmentation of the convergent margin and for plate tectonic models. J Geophys Res 106:19207–19220

Beheregaray LB, Ciofi C, Geist D, Gibbs JP, Caccone A, Powell JR (2003) Genes record a prehistoric volcano eruption in the Galapagos. Science 302(5642):75

Brilha J (2016) Inventory and quantitative assessment of geosites and geodiversity sites: a review. Geoheritage 8(2):119–134

Brocx M (2008) Geoheritage: from global perspectives to local principles for conservation and planning. Western Australian Museum, Welshpool, WA, p 175

Brocx M, Semeniuk V (2007) Geoheritage and geoconservation—history, definition, scope and scale. J Royal Soc W Aust 90(2):53–87

Christensen NI, Mooney WD (1995) Seismic velocity structure and composition of the continental crust: a global view. J Geophys Res 100(B6):9761–9788

Cleal CJ, Thomas BA (1995) Palaeozoic palaeobotany of great britain. Geological conservation review series 9. Chapman and Hall, London

Clift PD (2005) Sedimentary evidence for moderate mantle temperature anomalies associated with hotspot volcanism. GSA Spec Paper 388:279–287

Condie K (2001) Mantle plumes and their record in Earth history. Cambridge University Press, Cambridge, p 306

Cowie JW (1993) *Report*. World Heritage/Patrimonie Mondial (The International Convention for Conservation of Cultural and Natural Sites (Including Geology and Palaeobiology), Working Group on Geological and Palaeobiological Sites—A cooperative project of UNESCO, IUGS, IGCP & IUCN, Unpublished report, November 1993, 32 p

Darwin C (1844) Geological observations on the volcanic islands, visited during the voyage of the H.M.S. Beagle. Smith, Elder, and Co., London, 175 p

De Wever P, Alterio I, Egoroff G, Cornée A, Bobrowsky P, Collin G, Duranthon F, Hill W, Lalanne A, Page K (2015) Geoheritage, a national inventory in France. Geoheritage 7(3):205–247

Deitz RS (1961) Continent and ocean basin evolution by spreading of the sea floor. Nature 190:854–857

DeMets C, Gordon RG, Stein S, Argus DF (1987) A revised estimate of Pacific-North America motion and implications for western North America plate boundary zone tectonics. Geophy Res Let 14(9):911–914

Desmond A, Moore J (1991) Darwin. Penguin, London, pp 808

Detrick RS, Crough ST (1978) Island subsidence, hot spots, and lithospheric thinning. J Geophys Res 83(B3):1236–1244

Diaz S et al (2015) The IPBES conceptual framework—connecting nature and people. Curr Opi Environ Sustain 14:1–16

Dunkin EHW (1870) Anticipated destruction of the cheesewring. Nature 2(32):101. https://doi.org/10.1038/002101c0

Ellis NV, Bowen DQ, Campbell S, Knill JL, McKirdy AP, Prosser CP, Vincent MA, Wilson RCL (1996) An introduction to the geological conservation review. Geological Conservation Review Series No. 1, Joint Nature Conservation Committee, Peterborough, UK, 131 p

Epler B (2007) Tourism, the economy, population growth, and conservation in Galápagos. Charles Darwin Foundation

Fisher R, Hess HH (1963) Trenches, in the sea. In: Hill MN (ed) The Earth beneath the Sea: interscience. Wiley, New York 3: 411–436

Foulger G (2011) Plates versus plumes: a geological controversy. Wiley, Hoboken, NJ, p 364

Garcia-Cortés A (ed) (2009) Spanish geological framework and geosites: an approach to Spanisn geological heritage of international significance. Instituto Geologico y Minero de Espana, Madrid

Garcia-Cortés A, Gallego E, Carcavilla L (2012) Spain. In: Wimbledon WAP, Smith-Meyer S (eds) Geoheritage in europe and its conservation. ProGEO, Oslo, pp 334–343

Geist D, McBirney AR, Duncan RA (1986) Geology and petrogenesis of lavas from San Cristobal Island, Galapagos Archipelago. GSA Bull 97:555–566

Geist D (1996) On the emergence and submergence of the Galapagos Islands. Noticias de Galapagos 56:5–8

Geist D, Nauman TR, Standish JJ, Kurz MD, Harp KS, White WM, Fornari DJ (2005) Wolf Volcano, Galapagos Archipelago: melting and magmatic evolution at the margins of a mantle plume. J. Pet 46 (11):2197–2224

Geist D, Dieffenbach B, Fornari D, Curz M, Harpp K, Blusztajn (2008) Construction of the Galapagos platform by large submarin volcanic terraces. Geochem, Geophys, Geosyst 9. https://doi.org/10.1029/2007gc001795

Geist D, Harpp KS, Naumann TR, Poland M, Chadwick W (2008b) The 2005 eruption of Sierra Negra volcano, Galapagos. Ecuador Bull Volc 70:655–673

Geist DJ, Snell H, Snell H, Goddard C, Kurz MD (2014) A paleogeographic model of the Galapagos Islands and a biogeographical and evolutionary implications. In: Harpp KS, Mittelstaedt E, d'Ozouville N, Graham DW (eds) The Galapagos: a natural laboratory for the Earth Sciences. Wiley, Hoboken, pp 145–166

Global Volcanism Program (1991) Report on Marchena (Ecuador). In: McClelland L (ed) Bulletin of the global volcanism network 16:10. http://dx.doi.org/10.5479/si.GVP.BGVN199110-353080

Grant KT, Estes GB (2009) Darwin in Galapagos: footsteps to a new world. Princeton University Press, Princeton, NJ, p 362

Gray M (2004) Geodiversity. Valuing and conserving abiotic nature. Wiley, Chichester

Gray M (2013) Geodiversity: valuing and conserving abiotic nature (2nd edn). Wiley-Blackwell, 495 pp (1st edn 2003)

Gutscher MA, Malavieille J, Lallemand S, Collot JY (1999) Tectonic segmentation of the North Andean margin: impact of the Carnegie Ridge collision. Earth Planet Sci Lett 168(3–4):255–270

Harpp KS, White WM (2001) Tracing a mantle plume: isotopic and trace element variations of Galapagos seamounts. Geochem Geophys Geosyst 2:1042. https://doi.org/10.1029/2000gc000137

Harpp KS, Geist D (2002) The Wolf-Darwin lineament and plume-ridge interaction in the Northern Galapagos. Geochem Geophys Geosyst 3(11):8504. https://doi.org/10.1029/2002GC000370

Harpp KS, Hall PS, Jackson MG (2014) Galapagos and Easter: a tale of two hotspots. In: Harpp KS, Mittelstaedt E, d'Ozouville N, Graham DW (eds) The Galapagos: a natural laboratory for the Earth Sciences. Wiley, Hoboken, NJ, pp 145–166

Hillier JK, Watts AB (2005) Relationship between depth and age in the North Pacific Ocean. J Geophys Res 110:B2

Hernle K, van den Bogaard P, Werner R, Lissinna B, Hauff F, Alvarado G, Garbe-Schönberg D (2002) Missing history (16–71 Ma) of the Galapagos hotspot: implications for the tectonic and biological evolution of the Americas. Geology 30(9):795–798

Hess HH (1965) Mid-oceanic ridges and tectonics of the sea-floor 1965 Colston Symposium. Colston Papers Whittard WF, Bradshaw R (eds) Submarine Geol Geophys 17:317–334

Instituto Geofisico (2015) Informe Volcanico Especial Galapagos No. 2 Erupcion Volcan Wolf. www.igepn.edu.ec

Johansson CE (2000) Geodiversitet I Nordisk Naturvard. Nordisk Ministerraas, Copenhagen

Le Pichon X (1968) Sea-floor spreading and continental drift. J Geophys Res 73(12):3661–3697

Matthews TJ (2014) Integrating geoconservation and biodiversity conservation: theoretical foundations and conservation recommendations in a european context. Geoheritage 6:57–70

McBirney AR, Williams H (1969) Geology and petrology of the Galapagos Islands [needed]

Meschede M, Barckhausen U (2001) The relationship of the Cocos and Carnegie ridges: age constraints from paleogeographic reconstructions. Int J Earth Sci 90:386–392

Miller KG, Kominz MA, Browning JV, Wright JD, Mountain GS, Katz ME, Sugarman PJ, Cramer BS, Christie-Blick N, Pekar SF (2005) The Phanerozoic record of global sea-level change. Science 310(5752):1293–1298

Montelli R, Nolet G, Dahlen FA, Masters G, Engdahl ER, Hung SH (2004) Finite-frequency tomography reveals a variety of plumes in the mantle. Science 303(5656):338–343

Morgan WJ (1968) Rises, trenches, great faults, and crustal blocks. J Geophys Res 73(6):1959–1982

Müller RD, Sdrolias M, Gaina C, Roest WR (2008) Age, spreading rates, and spreading asymmetry of the world's ocean crust. Geochem Geophys Geosyst 9:4

NATURE CONSERVANCY COUNCIL (1990) Earth science conservation—a strategy. Nature Conservancy Council, Peterborough, pp 84

Nauman T, Geist D, Kurz M (2002) Petrology and geochemistry of Vocan Cerro Azul: petrologic diversity among the western Galapagos volcanoes. J Petrol 43(5):859–883

Németh K, Casadevall T, Moufti MR, Marti J (2017) Volcanic Geoheritage. In: Németh et al. Special Issue: volcanic geoheritage. Geoheritage 9(3):251–254

Nicholls H (2014) The galapagos: a natural history. Profile Books, pp 195

Page KN (2017) Integrating geodiversity and biodiversity conservation: strategic and site-based approaches. In: De Wever P et al (eds) Geoheritage inventories: challenges, achievements and perspectives, 22–26 September 2015. MNHN, Paris

Page KN (2018) Fossils, heritage and conservation: managing demands on a precious resource. In: Reynard E, Brilha J (eds) Geoheritage. Elsevier, Chennai, pp 107–128

Pereira D, Page K (2016) A new IUGS commission for geoheritage: the 'ICG'. Episodes 40(1):77–78

Porter DM (1985) The beagle collector and his collections. In: Kohn DH (ed) Darwinian Heritage. Princeton, UP, pp 973–1019

Powell RB, Ham SH (2008) Can ecotourism interpretation really lead to pro-Conservation knowledge, attitudes and behaviour? evidence from the galapagos islands. J Sustainable Tour 16(4):467–489

Quiroga D (2013) Changing views of the galápagos. In Walsh SJ, Mena CF (eds) Science and conservation in the galápagos islands; frameworks and perspectives. Springer, London, pp 23–48

Reynard E, Coratza P, Regolini-Bissig, G (eds) (2009) Geomorphosites. In: Pfeil V, Munchen Reynard E, Coratza P (eds) Geomorphosites and geodiversity: a new domain of research. Geogr Helv 62(3)

Reynard E, Fontana G, Kozlik L, Scapozza C (2007) A method for assessing "scientific" and "additional values" of geomorphosites. Geogr Helv 62(3):148–158

Reynard E, Pica A, Coratza P (2017) Urban geomorphological heritage. An overview. Quaestiones Geographicae 36(3):7–20. https://doi.org/10.1515/quageo-2017-0022

Reynard E, Perret A, Bussard J, Grangier L, Martin S (2016) Integrated approach for the inventory and management of geomorphological heritage at the regional scale. Geoheritage 8:43–60

Reynolds RW, Geist D, Kurz MD (1995) Physical volcanology and structural development of Sierra Negra volcano, Isabela Island Galapagos archipelago. GSA Bull 107(12):1398–1410

Sallarès V, Charvis PH (2003) Crustal thickness constraints on the geodynamic evolution of the Galapagos volcanic province. Earth Planet Sci Lett 214(3–4):545–559

Sharples C (2002) Concepts and principles of geoconservation. Parks and Wildlife Service, Tasmania (pdf document on website cited by Grey 2013)

SPANISH geological frameworks and geosites: an approach to Spanish geological heritage of international relevance. In García Cortés A, Águeda Villar J, Palacio Suárez-Valgrande J, Salvador González CI (eds), Instituto Geológico y Minero de España, Madrid 2009

Stanley, M (2001) Editorial. Geodiversity Update 1.1

Swanson FJ, Baitis HW, Lexa J, Dymond J (1974) Geology of santiago, rábida, and pinzón islands, galápagos. GSA Bull 85 (11):1803–1810

Teasdale R, Geist D, Kurz M, Harpp K (2005) 1998 Eruption at Volcán Cerro Azul, Galápagos Islands: I. Syn-Eruptive Petrogenesis. Bull Volcanol 67(2):170–185

Tefogoum GZ, Dongmo AK, Nkouathio DG, Wandji P, Dedzo MG (2014) Geomorphological features of the Manengouba Volcano (Cameroon Line): assets for potential geopark development. Geoheritage 6:225–239

Thomas BA, Warren LM (2008) Geological conservation in the nineteenth and early twentieth centuries. Geological Society, London, Special Publications 300(1):17–30

Thompson MME, Kempton PD, White RV, Kerr AC, Tarney J, Saunders AD, Fitton JG, McBirney A (2004) Hf-Nd Isotope constraints on the origin of the Cretaceous Caribbean plateayu and its relationship to the Galapagos plume. Earth Planet Sci Lett 217 (1–2):59–75

Titus SJ, DeMets C, Tikoff B (2005) New slip rate estimates for the creeping segment of the San andreas fault, California. Geology 33 (3):202–208

van Ree CCDF, Beukering PJH (2016) Geosystem services: a concept in support of sustainable development of the subsurface. Ecosyst Serv 20:30–36

van Wyk de Vries B, Matenco L, Gogus O, Byrne P, Einarson P, Delcamp A, Mossoux S, Hagos M, Nemeth K, Harangi S, Guilbaud M, Petronis MS, Rapprich V, Rose W, Vye E, Jerram D (2017) A global framework for the Earth: putting geological sciences in context. Global and Planetary Change. https://doi.org/10.1016/j.gloplacha.2017.12.019

Vine FJ, Matthews DH (1963) Magnetic anomalies over oceanic ridges. Nature 199:947–949

White WM, McBirney AR, Duncan RA (1993) Petrology and geochemistry of the Galapagos Islands: portrait of a pathological mantle plume. J Geophys Res 98:19533–19563

White RS, McKenzie D, O'Nions K (1995) Oceanic crustal thickness from seismic measurements and rare earth element inversions. J Geophys Res 97(B13):1

Wilson C (ed) (1994) Earth heritage conservation. Geological Society, London, and Open University, Milton Keynes (UK)

Wilson JT (1965) A new class of faults and their bearing on continental drift. Nature 207(4995):343–347

Wimbledon WAP, Smith-Meyer S (eds) (2012) Geoheritage in europe and its conservation. ProGEO, Oslo

Wimbledon WAP, Andersen S, Cleal CJ, Cowie JW, Erikstad L, Gonggrijp GP, Johansson CE, Karis LO, Suominen V (1999) Geological world heritage: GEOSITES a global comparative site inventory to enable prioritisation for conservation. Mem Descrit Carta Geol Ital 54:45–60

Wimbledon WAP, Ishchenko AA, Gerasimenko NP, Karis LO, Suominen V, Ohansson CE, Freden C (2000) Geosites—an IUGS initiative: science supported by conservation. In: Barettino D, Wimbledon WAP, Gallego E (eds) Geological heritage: its conservation and management. ITGE, Madrid, pp 69–94

Wimbledon WAP (1996) GEOSITES—a new IUGS initiative to compile a global comparative site inventory, an aid to international and national conservation activity. Episodes 19:87–88

Woo KS, Ju SO, Brilha (2018) Key geoheritage area: a potential new programme for geohertage conservation. In: Glowniak E, Wasilowska A, Leonowicz P (eds) Programme and abstract book, IX symposium ProGEO: geoheritage and geoconservation: modern approaches and applications towards the 2030 agenda, 25–28th June 2018, Checiny, Poland, pp 20

Worboys GL, Lockwood M, Kothari A, Feary S, Pulsford I (2015) Worboys GL, Lockwood M, Kothari A (eds) Protected area governance and management. Australian National University Press, Canberra

3 The Origins and Ecology of the Galapagos Islands

3.1 Introduction

Oceanic islands have a series of characteristics that make them unique for understanding biological evolutionary processes. One of the main characteristics of oceanic islands is that the distance between the mainland and the islands acts as an important biological filter. In the case of archipelagoes, the distance between islands is also an important factor that generates the conditions for the generation of speciation process in the case of many organisms. Many species of plants and animals are not able to survive the long-distance travel and of those organisms that are able to arrive, few succeed in becoming established on islands. This is a characteristic that explains the existence of certain biological processes that are typical of oceanic islands, including the founder effect, genetic drift, disharmonic biota, adaptive radiation, dwarfism and gigantism, ecological release, high endemism and rapid island evolution. These evolutionary processes can explain the existence, shapes and distribution of many of the organisms (Whitaker and Fernandez Palacios 2007). As described in Chap. 2, geological processes explain the formation and evolution of the islands, and these processes result in a constant emergence of empty biological niches as new land areas are created or others 'sterilized' by volcanic activity (e.g. lava flows and ash cover). These empty niches mean that the animals and plants that arrive must find their place and evolve under the new conditions.

The *founder effect* refers to the fact that in isolated geographical areas, such as oceanic islands, usually only small populations of any particular species are able to arrive, which represent only a small subset of the genes of the whole population (Whittaker and Fernandez Palacios 2007). Thus, through *genetic drift* this subset can produce novel and diverse forms as they start evolving to occupy the different open niches. The narrowing of the gene pool due to this restricted, or bottle-neck, effect can reshape the genetic base of the old adaptive system, thus providing new genetic variance for the new population (Whitaker and Fernandez Palacios 2007).

Disharmonic taxonomy refers to the fact that there is a distribution of organisms on Oceanic Islands that does not correspond to that on the nearer continental areas. Disharmonic taxa (i.e. arriving from climactically and ecological varied sources) provide evidence that helps us understand the process of dispersal, arrival, colonization and adaptation of the different species. In part, the disharmony (Whittaker and Fernandez Palacios 2007) is related to the fact that the climate and the biology of islands tends to be cooler than nearby continents, which is something we can see in the Galapagos, were penguins and sea lions live on an equatorial island. Explaining the differences in the composition of the fauna and flora between the mainland and the Galapagos requires understanding of the capacity of plants and animals to arrive, settle, reproduce and evolve. Some animals and plants do not have this capacity, as a result there are typically few, if any, amphibians, freshwater fish, dioecious plants, and mammals—other than flying mammals and maybe rats—on oceanic islands. This process changes through time as changes in distance and climate may have made it in the past either easier or more difficult for some groups of animals to arrive and colonize these remote islands. With a lower relative sea levels, as existed during Pleistocene glacial phases, it would also be easier for species to move from one island to the other (Grant and Grant 2008, pp 179–187; see Fig. 2.11 in Chap. 2).

Another common biological process in oceanic islands is *ecological release*, resulting from the freeing of limiting factors on the environment. This occurs when species arrive to an island where they do not encounter many of the competitors and predators that characterize continental ecosystems. In some cases, this can lead to *adaptive radiation*, an increase of the population and rapid evolutionary processes (Kohn 1972). Radiation occurs more often in remote and high islands were there are many available (e.g. vacant) ecological niches. Different species have diverse radiation zones (MacArthur and Wilson 1967), and it is in the edge of these radiation where often greater radiation may occur (Paulay 1994, 134–144); Whittaker and Fernandez

D. Kelley et al., *In the Footsteps of Darwin: Geoheritage, Geotourism and Conservation in the Galapagos Islands*, Geoheritage, Geoparks and Geotourism, https://doi.org/10.1007/978-3-030-05915-6_3

Palacios 2007). Inevitably, competition for resources may lead to *dwarfism* when compared to originator species; a good example of insular dwarfism specially found among large mammals such as the dwarf elephants of the Mediterranean and Indonesia and the extant pigmy Asian hippopotamus that used to be found in Java and the extinct Madagascar and Mediterranean hippopotamus (Poulakakis et al. 2002; Weston and Lister 2009; Van Den Bergh et al. 2008). Conversely, gigantism is often a feature of island species where an absence of typical mainland predators, may allow some to grow to a larger size, the Galapagos tortoises being a dubious example (Raia and Meiri 2006; Caccone et al. 1999). Animals and plants with a restricted—sometimes very restricted—geographical range can be described as *endemic* to that area.

Adaptive radiation refers to the evolution of distinct species and varieties from an ancestral form. The availability of empty niches is a key condition that allows species to diversify. The Hawaiian honeycreepers and the Galapagos finches, among birds; land snails, among invertebrates; and the Asteracea, among plant species, are often cited as representative examples of the process (Whittaker and Fernandez Palacios 2007). In Hawaii, around 980–1000 species of plants arose from possibly 270 original colonists (Wagner and Funk 1995). There are an estimated 10,000 insects in Hawaii that may have evolved from some 350 to 400 arriving species, crickets and drosophilids being well known examples of insects that underwent adaptive radiation across these islands (Paulay 1994, 134–144; Wagener and Funk 1995; Whittaker and Fernandez 2007). As we will see a similar process has occurred in the Galapagos.

3.2 Terrestrial Flora

When Darwin arrived to the Galapagos he referred to the flora in rather negative terms, as "*wretched-looking little weeds*" (Darwin 1845). He soon, however, began to recognize similarities of many of the species with those from mainland South America, which he had observed earlier during the voyage of the Beagle. Indeed, it was the plants which seem to have first caught his attention and started him thinking about the origins of the Galapogean biota (Nichols 1997). Darwin collected some 210 species of plants and commented on the possible origin of the plants in the Galapagos. He also sought advice from his mentor at Cambridge, John Steven Henslow (Walters and Stow 2002) and from J. D. Hooker who examined most of the plants. Hooker described 78 of them as being new species and analyzed the close relationship of them to the plants of the South American continent. The fact that more than 50% of the species where not found anywhere was important for the development of the idea that plants as well as animals had evolved from their mainland ancestors. Hooker characterized Asteraceae as the most unique of the family of the Galapagos Islands (Stocklin 2009) (Fig. 3.1).

One of the most important modern reviews of the flora of the Galapagos is that of Porter (1983), who described the possible way in which the different species of plants arrived to the islands. Of the 543 indigenous species, Porter calculated that 231 were endemic, that the current indigenous taxa came from 378 ancestors, and the vascular plants came mostly from South America (99%) and only 1% from North and Central America (pp. 36–96).

Plants found in oceanic islands have a capacity for long-range dispersal. This is often a characteristic of their seeds and propagules, for example their capacity to be taken by the wind, to survive in the gut of birds or to become attached to the feathers, fur and other parts of animals (McMullen 1990a, b). Others may also have a resistance to long-exposure to sea-water and arrive though drifting. Examples of plants that arrived by sea in this way include Red Mangrove (*Rhizophora mangle* (L.)), Sea Purslane (*Sesuvium portulacastrum* (L.)), Common Purslane (*Portulaca oleracea* (L.)), Beach Morning Glory (*Ipomoea pes-caprae* (L.)), and the Beach Bean (*Canavalia rosea* (Sw.)).

It is now thought that 236, or 43%, of the indigenous taxa (i.e. 236 of 543 species) are endemic (Tye and Ortega 2011). Porter (1983) considered there are 436 species that come from 306 introductions. Of the plants that came carried by birds, Porter considers that 59% where introduced by birds, 32% by wind and the rest, 9%, by the sea. Of the proportion that were carried by birds 64% carried in the digestive tract, 21% in the feathers and 15% on mud attached to legs (pp. 36–96). In a more recent study, it was concluded that endozoochory (animal digestion) accounted for 16.4% of the flora, epizoochory (animal adhesion), 15.7%, hydrozochory (sea water), 18.6%, anemochory (wind) 13.3% and 'unassisted', 36% (Vargas et al. 2012).

South America is thought to have contributed to 45% of the colonization, the other sources are Central America and the Caribbean at 12%, and North America at 5% (Tye and Ortega 2011). This relatively strong connection with Central America is probably explained by the geological evidence linking the Galapagos hotspot to that region (see Chap. 2), perhaps suggesting that a certain amount of 'island-hoping' has occurred (i.e. species crossing the relatively short distances between successively developed island in the chain, most of which are now submerged. In addition, before the final closure of the Panama Isthmus 2.7 Ma, there would have been more maritime connection with the Caribbean (Haug and Tiedemann 1998).

Porter (pp. 36–96), also maintained that evolutionary radiation occurred in the arid zone which he thought had not changed during the Holocene and that preceding periods had been dryer (Tye and Ortega 2011). New paleobotanical

Fig. 3.1 A typical Galapagos, Daisy-Tree, or *Scalesia*, in the botanical gardens of the Charles Darwin Research Center—height is around 2.5 m. *Photo* K. N. Page

research is also now confirming that many species that were once thought to be introduced are actually native to the Galapagos, such as, *Hibiscus diversifolius* Jacq. and *Ageratum conyzoides* L. (Tye and Ortega 2011). There are 19 plant genera that are endemic to the Galapagos and many endemic species. The latter include the 6 species and 14 varieties of the prickly pear cactus, *Opuntia*, including unique arborescent forms, up to 12 m height. The latter developed to avoid grazing by giant tortoises, some of which can stretch their heads to a height of around 1.7 m to reach more tender parts of the plant (Barthlott and Porembski 1994). There are also 14 species and 20 of Alternanthera—the 'Chaff Flower'—, and 15 species and 19 varieties of *Scalesia* (Daisy Tree)—all being o examples of adaptive radiation. Orchids on the Galapágos provide a good example of a disharmonic biota, with more than four thousand species found in mainland Ecuador, but only 14 across the islands. As well as issues concerning long-distance transport, many orchids are pollinated by birds, mammals and insects that are not found in the Galapagos, and hence even if they had arrived, they would have been unable to reproduce.

The total flora of the Galapagos comprises some 749 angiosperm species and 192 genera, with 216 (28.8%) of the former and 7 (4%) of the latter being endemic (McMullen 1999). The majority of these plants are self-compatible (McMullen 1999) and most are autogamous, making it possible for them to reproduce without the aid of pollinators, a characteristic often found with plants of oceanic islands (McMullen 1999).

3.2.1 Floristic Zones

The classic distribution of different floristic zones across the islands is a reflection of variations of the ecology of the different islands which are strongly linked to the age and altitude of each island. Thus, in those islands that have

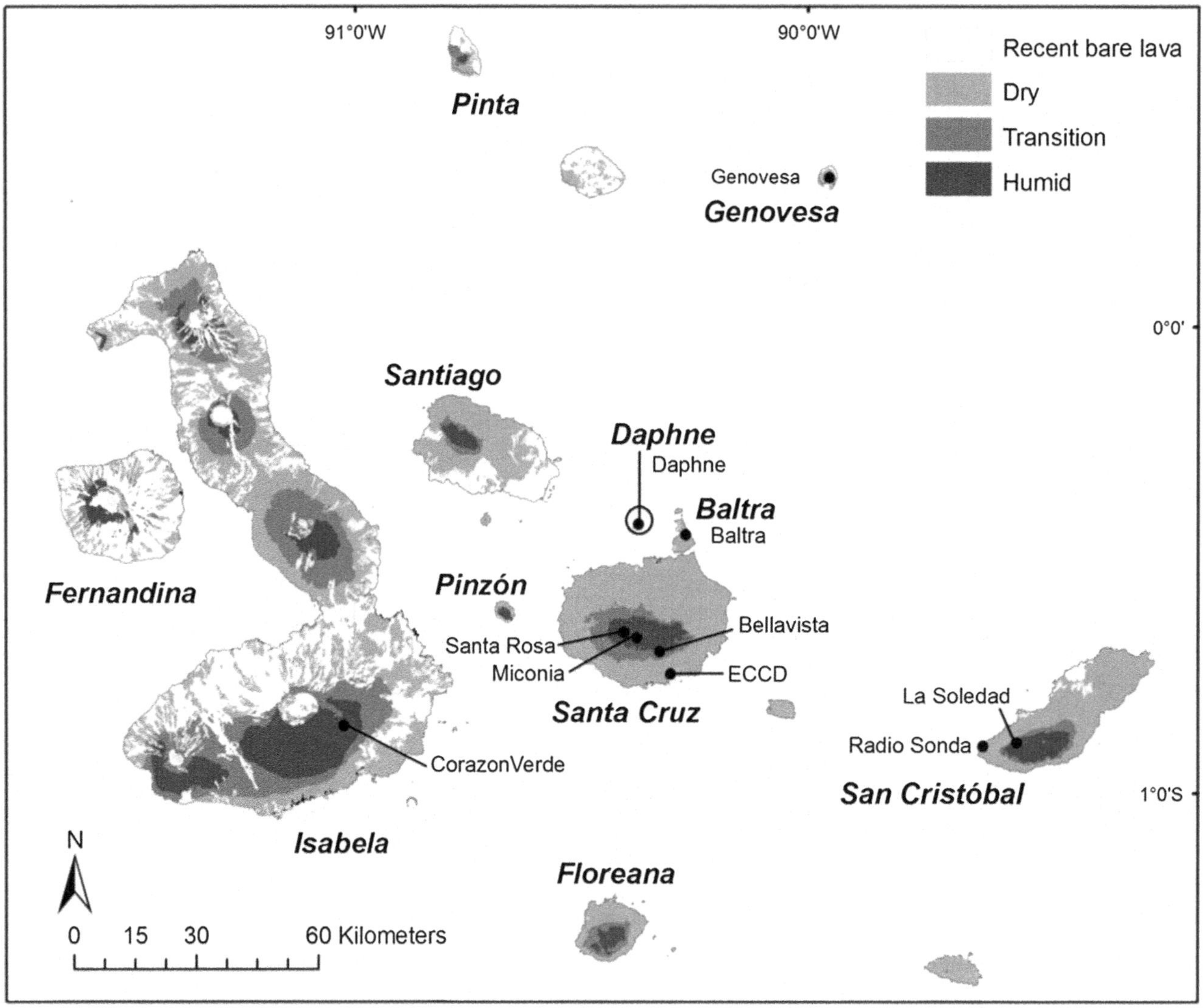

Fig. 3.2 Vegetation zones across the Galapagos: adapted from Trueman and d'Ozouville (2010) after Ingala and Orstom (1989)

grown to a higher elevation as the result of long term volcanic activity, natural condensation leads to the development of clouds that create the conditions for a humid zone. This phenomenon was often remarked upon by earlier visitors, leading to the Galapagos often being referred to as the 'Enchanted Islands'. These humid conditions have led, on the older islands, to the development of relative rich tropical soils due to weathering of the volcanic bedrock, thus creating very different growing conditions and floras to the dry lowlands—but inevitably attracting farming settlers… Flatter islands that cannot trap clouds have a more limited range of floristic zones, and are dryer thus characterized by xerophytic species (Wiggins and Porter 1971; Tye and Ortega 2011).

Seasonality, as well as elevation, also plays an important role in the distribution of plants and animals. There are two distinct seasons in the Galapagos, during the cool season the Humboldt Current is prevalent bringing cold water northwards from the west coast of South America, resulting in cooler air temperatures and what is called by the local people the *guarua* season. The warm season from January to May is characterized by a warming of the water, less wind and occasional showers (Fig. 3.2).

Moving from sea-level to the highest elevation of the largest islands, a number of floristic zones are encountered:

(a) **Littoral Zone**: The lowest, closest to the shore is the *littoral zone* which is characterized by plants adapted to salty, coastal conditions such as saltwort (*Batis maritima*), different types of mangrove (Black Mangrove, *Avicennia germinans* (L.); White Mangrove, *Laguncularia racemose* (L.); Red Mangrove, *Rhizophora mangle*; and Button Mangrove—or Buttonwood, *Conocarpus erecta* L.), Galapagos Lycium (*Lycium minimum* C.L. Hitc.), and the endemic Galápagos Carpetweed (*Sesuvium edmonstonei* Hook) (Wiggins and Porter 1971; McMullen 1999).

Fig. 3.3 The Littoral Zone on Sombrero Chino ('Chinese Hat'), dominated by dessication-resitant and salt tolerant plants. *Photos* K. N. Page

It is this zone where most of the infrastructure related to settlements, ports, hotels, roads and other buildings have been placed, but it is also the zone that generates many of the ecosystem services for tourism, as many of the established paths for tourists are found in this and the arid zone (Fig. 3.4). Indeed, as most of the tourism in Galapagos is conducted on ships and motor boats, and this zone is closest to the ocean, it is often the only area that tourists visit on many of the islands. Nevertheless, in addition to localized human interference, this is also the zone most affected by natural processes such as ongoing coastal geomorphological changes, including sea-level changes and sedimentary regimes (e.g. the evolution of sand beaches and associated features, and lagoonal areas of mud deposition) and the morphology of rocky coastlines, including cliffs (Figs. 3.3 and 3.4).

(b) **Arid Zone**: Following the littoral zone is the *arid zone*, the most extensive of all zones across the islands. According to Wiggins and Porter (1971), McMullen (1999), this zone reaches an altitude of 120 m on the southern slopes and 300 m on the northern side of the islands (reflecting the prevailing wind conditions). In this zone, many of the plants are deciduous, to aid survival through the dry season. Herbaceous species which arrive on islands where there are no trees, may subsequently have to compete with other vegetation by developing tree-like forms, which often leads to the evolution of tree-like plants on islands when this type of morphology was never present in ancestral mainland communities (Losos and Rickefs 2009).

In the Galapagos, a good example of this phenomenon is the development of tree-like forms of the Prickly Pear Cactus, *Opuntia*,. The large number of species and varieties of genus across the arid zones of the islands and is also a very good example of adaptive radiation, as many of the different forms are restricted to specific islands (McMullen 1999). Some of these species, for instance *Opuntia echios* J. T. Howell, which can reach 12 m in height, exemplify a phenomenon that can be found on other islands across the world, where some non-woody plants, start to develop structures that resemble those of the woody plants, including a bark (Fig. 3.5) (Whittaker and Fernandez Palacios 2007) (Fig. 3.6).

Other trees and tree-like forms that are adapted to these dry environments with only seasonal rains include: Matazamo, (*Piscidia carthagenensis* Jacq.), Galapagos acacia (*Acacia rorudiana* Cristoph.), 'palo santo' or Incense Tree, (*Bursera graveolens* (Kunth)) and bushes such as Galapagos

Fig. 3.4 Mangroves in Bahia Elizabeth (Elisabeth Bay), the typical coloniser of sheltered embayments. *Photo* K. N. Page

croton (*Croton scouleri* Hook.), Darwin's cotton (*Gossypium darwinii* Watt.), and Parkinsonia (*Parkinsonia aculeate* L.) (Wiggins and Porter 1971; Porter 1983; McMullen 1999). Unlike higher and more humid areas, the soils of the arid zone are typically thin and poor due to the relatively limited weathering of the volcanic bedrock.

(c) **Transition Zone**: The *transition zone* reaches up to approximately 200 m of elevation or higher (McCullen 1999). This zone is composed of both herbaceous and evergreen trees including the Incense Tree, Yellow Cordia (*Cordia lutea* Lam.), Pega Pega (or Galapagos Pisonia) (*Pisonia floribunda* Hook.) and Guayabillo, or Galapagos, Guava (*Psidium galapageium* Hook.), (Wiggins and Porter 1971; McMullen 1999). As it names suggest, this zone represents a transistion between arid regions and higher, more humid regions with evergreen trees beginning to appear due to a less extreme seasonal aridity (Fig. 3.7).

(d) **Scalesia Zone**: The next zone in increasing elevation is the *Scalesia zone,* typically ranging from 400 to 500 m above sea level. The 'Daisy Tree' *Scalesia*, including *Scalesia pedunculata* Hook. dominates this zone. The zone is very humid and the soils are the richest across the islands, meaning that on the inhabited islands, this zone has been the most disturbed by agriculture. During the cool season, the Scalesia forests are characterised by typical *garua* mist (or thick fog) which provides moisture for the trees, but also for many epiphytic plants, mosses, ferns and liverworts. Other types of shrubs characteristic of this zone include Galapagos Guava and Cat's Claw (*Zanthoxylum fagara* (L.)) and epiphytic plants including orchids and lower plants such as mosses and ferns start to become common (McMullen 1999) (Fig. 3.8).

(e) **Zanthoxylum or 'Brown' Zone**: The *Scalesia* zone is followed by the 'Brown Zone' were where the evergreen 'cats claws' tree *Zanthoxylum* fagara dominates along with White-haired Tournefortia (*Tournefortia pubescens* Hook.) and Galapagos Acnistus (Acnistus ellipticus Hook.) dominate. Trees and epiphytes are common in this area which has been described as the brown zone due to the large number of lichens that cover the trees (McMullen 1999). This is another of the floral zones most affect by the agricultural activities across the islands and many invasive species have become established, as introduced by farmers (Walsh et al. 2008) (Fig. 3.9).

Fig. 3.5 The Arid Zone on Isla Rabida in July (dry season)—leafless 'Palo Santo' trees cloak the hill side with tree-like *Opuntia* at the cliff top. *Photo* K. N. Page

(f) **Miconia Zone**: The next higher zone is the *Miconia Zone*, which begins at around 400 m on San Cristobal and Santa Cruz Islands, and is characterized by the presence almost exclusively of the shrub *Miconia robinsoniana* Cogn. (McMullen 1999). This zone, along with the other higher zones, have been affected by the conversion of the natural vegetation into pasture and agriculture. The greater abundance of water and the fact that the soils are more fertile has also meant that the larger *fincas*, or farming estates, have become established in these higher zones. Many of the agricultural plants introduced into such areas have spread and became invasive species, including Common Guava (*Psidium guayava* L.), Curse of India (*Lantana camara* L.), Hill Raspberry *(Rubus niveus* Thunb.) and Quinine (*Cinchona pubesceris/ succirubra* Pav.). On some of the Island such as Floreana, San Cristobal and Isabela these changes started in the 19th century with the establishment of the first colonists, but in the case of Quinine, the first recorded introduction was as late as 1946 (McMullen 1999).

(g) **Pampa Zone**: The next higher zone is the 'fern sedge' or Pampa Zone, which typically begins at an elevation of around 550 m (Wiggins and Porter 1971; Porter 1983; McMullen 1999). The Pampa Zone is found at the highest levels of the highest islands such as Santa Cruz and Isabella, and consists mostly of ferns, sedges and grasses—hence it is sometimes called the 'fern-sedge zone' (e.g. See Fig. 3.11). Different species of ferns grow in this region, including Pteridium aquilinum and the endemic tree fern Cyathea weatherbyana Galapagos tree fern (Cyatheaceae) which can reach 3 m. During the cool season this zone is almost always wet, with pools of water and the vegetation making it an ideal zone for tortoises. This zone is also threatened by invasive plants such as the quinine tree in the past the presence of goats and pigs (McMullen 1999) (Fig. 3.10).

There is only one permanent fresh water lake in the Galapagos which is found in San Cristobal but in the higher zones many fresh water springs can be found in the tallest islands. The porous nature of the soil in some of the Islands has meant that most of the year there is few surface water in most of the islands.

Locally, these higher zones are also visited by significant numbers of tourists, as many lie outwith the GNP and hence access is less restricted—especially where former agricultural fincas have developed touristic facilities and activities.

Fig. 3.6 **a** A typical tree-like *Opuntia echios* in the botanical gardens of the Charles Darwin Research Center—height is around 5–6 m. Note the development of 'bark'. *Photo* K. N. Page **b** *Opuntia echios* with a 'barky trunk' protecting it from the hungry tortoise. *Photo* D. F. Kelley

Key features visited include some of the larger lava tunnels, fresh water lakes such as el Junco (see Chap. 5), and encounters with giant tortoises and other members of the fauna that characterizes this region are an important attraction (Fig. 3.12).

3.2.2 Origins of the Endemic Flora of the Galapagos

Elements of the famous endemic flora are found throughout the Galapagos Islands, in different proportions in different floral zones. In the arid zone, 67% of the plants are endemic; in the humid zone, 29% of the plants are endemic; but only 4% of the vascular plants are endemic on the littoral zone (Porter 1983). There are various reasons for these differences in the distribution of endemic species across the floral zones. Coastal xerophytic species, for instance, enjoy greater opportunities for dispersal through oceanic currents. The arid and xerophytic habitats are the oldest habitat type across the islands as the earliest colonizers first arrived in these areas from the shore, hence the plants present have had more time to evolve. In addition, as most human arrivals colonized the *Miconia*, pampa and Transition zones, most of the introduced and invasive plants are found in these areas, which is why much of these habitats have been transformed.

The presence of the littoral and the arid zones in the islands acts as a second biological filter for many species after the ocean and the air. Thus, many species that came from the humid northern coast of the mainland of Ecuador and the Pacific coast of Colombia, e.g. the Choco region of Western Colombia and Northwestern Ecuador, had no easy way to reach to the higher, wetter regions of the islands where they could become established. Many may well have arrived to the islands floating on rafts of vegetation, especially during dramatic El Niño events, but would not then have been able to become established because they were not well suited to survive the harsh conditions of the lower dry areas.

Such areas, however, with their less well developed soils due to less weathering than can take place in wetter zones, do have a higher percentage of endemic plants. This could be in part be due to the longer persistence of this zone over time, which provided the conditions for various plant species

Fig. 3.7 Transition zone flora on Isabela Island. *Photo* D. F. Kelley

to differentiate from their mainland relatives. Nevertheless, it is also important to understand what happened within these zones in the past, as the climate was different during much of the Pleistocene (Grant and Grant 2008). In reality, it is likely that many of the plants that characterize the dry lower zones, arrived from regions which were also dryer during the Pleistocene, such as the southern part of South America, Central and south-eastern Ecuador, and Peru.

Scalesia and other Asteraceae are some of the best-known examples of adaptive radiation among plant species in Galapagos. The incredible diversity achieved by *Scalesia* indicates that there was a rapid process of evolution and adaption to the new niches that opened through the processes of geological formation and transformation of the islands. In particular, as islands built up to higher levels, the resulting development of relatively humid areas, lead to weathering of the volcanic bedrock and hence extensively soil formation. Another endemic Asteraceae is *Lecocarpus*, an endemic genus with three species, only found in the Galapagos. Other examples of radiation and adaptation among plants include the cacti, most notably *Opuntia* as mentioned previously, especially on those islands where iguanas and the tortoises pose a selective pressures on these plants, causing them to grow large and tree-like.

In contrast, endemism in littoral areas is more limited, as the ocean does not constitute a geographic barrier to the types of plant that inhabit such areas, as many of these plants, such as the mangroves, would have arrived in the islands by floating—hence they are not completely genetically isolated from other populations of the same species elsewhere. Highland plants, such as ferns, also have relatively limited levels of endemism as their spores are easily dispersed by the wind, including over large distances.

The flora of the Galapagos is disharmonic in composition because only certain types of plants have long-distance dispersal abilities and capacities to establish and adapt to the new conditions—they therefore tend to dominate those with poor abilities. Some examples of the species with the greater dispersal capacity are ferns with 38 genera and 100 species, Amaranthaceae (Amaranths) with 7 genera and 29 species,

Fig. 3.8 Vegetation in the area of Los Gemelos in Santa Cruz; showing both native *Scalesia* and invasive species. *Photo* D. F. Kelley

Cyperacea (sedges) with 7 genera and 35 species, and Poacea (grasses) with 33 genera and 65 species. The seeds of some of these plants can pass through the gut of birds and remain viable, or grow in muddy areas where they easily become attached to the legs of birds. Asteraceae (Daisys, etc) (38 genera and 65 species) have fruits with capillary bristles that allow them to be dispersed by the wind or barbs that help them become attached to flying animals such as marine birds (McMullen 1999). The seeds that arrive in the arid lowlands do better if they have longer periods of dormancy before germination, hence they can survive the long periods of drought of the zone. Dormancy also helps plants withstand long periods in the intestines of birds—the plants that do well in moist areas, however, tend to germinate immediately (Whittaker and Fernandez-Palacios 2007).

As in other oceanic islands, the lack of animal pollinators often results in greater amounts of plants that are *self-compatible* and *autogamous* (McMullen 1999) meaning that they have the capacity to fertilize themselves. In the Galapagos, there are few of the classical pollinators found in tropical areas, such as humming birds and some bat species, and wind does not play an important role in the pollination of many of the island's plants (here are two species of bats present, however, but they do not act as pollinators). There are, nonetheless, insects available for polination, the most important being the endemic carpenter bee *Xylocopa darwini* (McMullen 1987, 1989a, b), and it has been estimated that some 80 different plants species can be visited by the bee (McMullen 1999). The sulfur butterfly (*Phoebis sennae*) and the Galapagos blue butterfly (*Leptotes parrhasioides*) are also important pollinators (McMullen 1989a, 1990a, b, 1999) and nocturnal pollination by moths is important for some plants (Traveset et al. 2013). Some birds such as finches and mockingbirds have also been seen at the flowers of different plants, visiting for nectar, insects and pollen (Grant and Grant 2008; McMullen 1987, 1989b; Tye and Ortega 2011) and may also be important pollinators.

Recent studies have shown that the evolution of the Galapagos flora is more complex than previously thought (Tye and Ortega 2011). The dull colors, mainly yellows and whites, along with the small size of the flowers in the Galapagos can be explained by the relative scarcity of pollinators. Wind seems not to play a key role in the pollination process as structures that may promote wind pollination are largely absent (McMullen 1999). The Galapagos flora has a high proportion of self-compatible species and no evidence of the evolution of outcrossing mechanisms (1986), although the proportion of dioecious species (with male and female reproductive systems on separate plants), and hermaphroditic species (with both male, pollen-producing and female, ovule-producing parts in each flower) seems to be similar to those of the mainland of Ecuador, (McMullen 1989a, b; Tye and Ortega 2011). Other studies have concluded that the selection for autogamous, plants that can fertilize themselves

Fig. 3.9 *Zanthoxylum* Zone seen along path near Cueva de Sucre site on the flanks of Sierra Negra Volcano on Isabela Island. *Photo* D. F. Kelley

is less severe than was previously thought and it could be that the initial strategy with mainly autogamous species, changes with time as species becomes well established (Tye and Ortega 2011). Thus, a selection pressure for more outcrossing [i.e. self-infertile] would develop, as there is an increased fitness among plants with a small population size, if genetic variety is increased.

3.3 Terrestrial Fauna

In a similar way to the plants, the fauna that is able to arrive and become established in oceanic islands has special characteristics. As with the plants, where the production of spores, certain types of seeds and propagules provide species

Fig. 3.10 Miconia plants on Santa Cruz Island. *Photo* D. F. Kelley

with a higher dispersal ability (i.e. *vagility*), animals with certain types of larva or eggs, for instance with a greater potential to be transported by wind, other animals or ocean currents, are the most likely colonizers of oceanic islands. The successful arrival of such animals and plants, however, is also dependent on their capacity to withstand the long journey to the islands, either passively (e.g. floating) or actively (e.g. by flying as for birds and some insects).

Larger pelagic oceanic birds can fly long distances to islands and may also carry with them plant seeds, insects and microorganisms. Other animals, such as penguins and marine mammals, can arrive by swimming—this is active dispersal. Wind and the storms, however, are important mechanisms through which other animals can arrive on oceanic islands, in a more passive way, for instance carried by oceanic currents, either on vegetation rafts or floating as individuals or groups. Many reptiles, such as lizards and snakes, as well as small mammals such as rats, small birds and invertebrates such as snails and arthropods, may have arrived in this way (Whittaker and Fernandez Palacios 2007).

In an analogous way, the larva of many benthic (i.e. bottom-living) marine animals such as echinoderms and mollusks, as well as reef fish, will also have arrived carried

Fig. 3.11 Pampa Zone atop Sierra Negra volcano on Isabel Island. *Photo* D. F. Kelley

by ocean currents—but this would be a 'normal' distribution strategy, quite unlike the chance events that carried terrestrial animals. Once animals arrived in the islands, their ability to establish and successfully colonize, will depend, among other factors, on their reproductive and demographic viability (Whittaker and Fernandez Palacios 2007). Many animals require a mate to reproduce and arriving alone would mean a failure to become established—unless, of course, there were later arrivals of the same species after another event. It is not surprising, therefore, that the study of the process of establishment and subsequent radiation of vertebrate and invertebrate species across the Galapagos, has provided many classic examples of evolutionary processes.

It is common on oceanic island, for animals and plants to follow the 'island progression rule', as they occupy first the older islands and they proceed to colonize the younger islands, as niches are created through the process of island evolution—and in the Galapagos this is fundamentally geological. This rule has been proved in many other oceanic islands, including through the evolution of both vertebrates and invertebrates in Hawaii (Fleischer et al. 1998, 533–545; Juan et al. 2000; Wagner and Funk 1995). In the case of the Galapagos, such a process implies an evolutionary sequence from southeast to northwest, i.e. from the oldest to the youngest islands, with land snails (Chambers 1991), Galapagos tortoises (Caccone et al. 1999, 2002, 2004), and the Galapagos lava lizards (Kizirian et al. 2004, 761–769), amongst other animals, seeming to follow this rule.

Invertebrates provide important examples of the evolutionary processes occurring in islands and in the Galapagos, land snails and arthropods have evolved in interesting and surprising manners. In the case of land snails, there are 88 described species, of which the genus *Bulimulus* constitutes 80% of the diversity (Chambers 1991; Parent and Crespi 2009)—indeed *Bulimulus* has the richest variation of any genus of plant or animal in the archipelago (Parent and Crespi 2009). The Galapogean forms seem to be most closely related to continental South American bulimilids (Breure 1979),

Fig. 3.12 The El Mazanillo 'tortoise ranch' on Isla Santa Cruz—giant tortoises in an agriculturally altered semi-natural habitat. *Photo* J. Bello-Page

although Parent and Crespi (2006, 2009), Parent et al. (2008), consider that it is still risky to propose a single continental ancestor, as little is known about the distribution of groups across the mainland. *Bulimulus* has colonized almost all of the major islands, following roughly the sequence of their geological formation (Parent et al. 2006, 2008). Successful colonization was followed by independent radiation within each island (Parent and Crespi 2006) and genetic studies have indicated that many species are single island endemics. Species richness is also correlated with island area, maximum elevation, habitat diversity and geological age (Parent and Crespi 2006). *Bulimulis* is found in all the different vegetation zones except amongst lava boulders and sandy coastal areas (Parent and Crespi 2006, 2009).

Younger Islands such as Fernandina and Isabela have fewer land snail species than might be expected based on their areas and elevations, probably reflecting their age as diversity is higher on older but similarly sized islands such as San Cristobal (Parent and Crespi 2006, 2009). There is some evidence of adaptation to the different climactic-floral zones including as shell shape appears to be linked to moisture variations and elevation—thus species with more slender shells are found in the lower elevations and species that are more conical in shape are found in the more humid, higher elevations (Parent and Crespi 2006). Indeed, the diversity of niches predicts the number of within-island speciation events better than island area (Parent and Crespi 2006). Some of these species occur in sympatry, two species living in the same area without one displacing the other, and could have adapted to the different habitats where they occur. In the central islands, such as Isabela, Pinzon, Santa Cruz and Santiago, land snail assemblages that are polyphyletic have been found, whereas the more isolated islands such as San Cristobal, Floreana and Española, tend to have monophyletic assemblages resulting from within island diversification (Parent et al. 2008). Besides island size and niche diversity, island age also plays an important role in the diversity of species, as in older islands snails have had time to diversify more (Fig. 3.13).

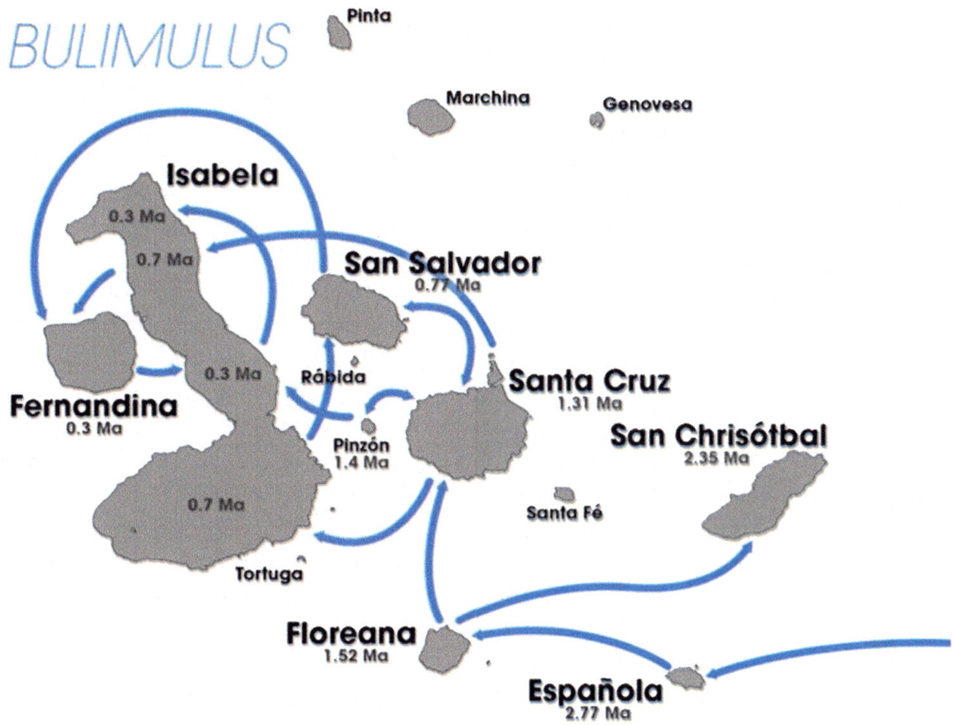

Fig. 3.13 The step by step colonisation of the Galapagos by the land snail *Bulimulus*, as modelled by Parent and Crespi (2006): From the first arrival on the oldest island Espanola to colonisation of the youngest, Fernandina is envisaged to have happened over a period of around 2.5 My. The shells of species of *Bulimulis* range from around 20 to 50 mm in length (from https://commons.wikimedia.org/wiki/File:Bulimulis_Colonization_of_Galapagos.png)

Surprisingly, there is not much information about the evolution of insects across the Islands. Very few insect lineages appear to have diversified, and only some 5% of the 1,000 or so genera recorded have undergone a process of speciation (Peck 2006). In the case of the Galapagos beetles (Coleoptera), some species split from their closest living relatives on the mainland over 7 Ma ago, so must have first colonised pre-existing islands, as the oldest today, Fernandina, is less than 3 million years old (Sequeira et al. 2008)—but it is also clear that there must have been several different colonization events. Beetle diversification has occurred between islands and within islands. On San Cristobal, for instance, one of the older islands and one which has a diversity of ecosystems, there has been perhaps the greatest diversification, although Santa Cruz may be similar (Sequeira et al. 2008).

In the case of the Galapagos Lepidoptera, diversification is thought to be begun some 3.3 Ma (Schmitz et al. 2007) and the group demonstrates that within island speciation occurs in taxa with low vagility (i.e. less mobile) and in large Islands within diverse environments. For instance, on Isabela, which has a series of relatively isolated volcanoes separated by recent lava flows, the scenario is very conducive to within-island speciation. In the case of other species that are more vagile, speciation between islands is more likely. For instance, *Galagete*, a genus of microlepidotera, which includes 12 species endemic to the Galapagos, represents the largest endemic radiation of the Lepidoptera in the archipelago. These 12 species have a variable distribution, with some encompassing the whole archipelago, while others have a single island distribution (Landry 2002). These species seem to be the product of a single colonization event, arriving from the coast of South America probably by passive aerial transport for instance carried by trade winds (Schmitz et al. 2007). There seem to have been two radiation events, one some 3.3 Ma ago and the other around 1.5–2.0 Ma—probably corresponding to the emerging of different islands and the creation of new, open niches for the species (Schmitz et al. 2007). Some of the younger islands have a large number of species, as is the case for Isabella and Fernandina that have eight species each (Schmitz et al. 2007), and hence, as for birds, the number of species in each island does not follow the island dispersal rule as for reptiles and land snails, but can be explained by the high mobility of flying species (Schmitz et al. 2007).

Thus, as in the case of plants, different types of animals differ in their capacity to arrive and to become established in the islands. Reptiles for example, with their low metabolic rates and relative tolerance of heat and lack of fresh water, are much better adapted to survive the long oceanic journey, whereas amphibians and mammals in general cannot survive long periods floating on the ocean. Amphibians in particular, would also have a hard time trying to reproduce in the dry zone, which is the area they would have encountered first upon arrival. In contrast, the hard, drought resistant skin, and hard shell of their eggs, would have allowed reptiles to survive the voyage and also to reproduce in the drier areas of the islands. In some cases, such as with geckos (as in some

Fig. 3.14 A widespread oceanic bird species, the Blue Footed Booby pictured in nesting site, Cabo Rosa, Isla San Cristobal. *Photo* D. F. Kelley

insects), through *parthenogenesis*, a single female can reproduce even in the absence of a male (Whittaker and Fernandez Palacios 2007). No native amphibians are found in the Galapagos and the only terrestrial mammals found are several species of rice rats (as discussed below).

Not surprisingly, birds are often the first animals to arrive to oceanic islands as large migratory species and those that can fly long distances can become established in islands relatively easily. Typical species which would have arrived in this way to the Galapagos include the Waved Albatross (*Phoebastria irrorata*), different species of boobies (e.g. Blue Footed, *Sula nebouxii*, Nazca, *Sula grantii*, and Red Footed, *Sula sula*), Greater Flamingo (*Phoenicopterus ruber*), Brown Pelican (*Pelecanus occidentalis*) and the two species of frigate birds, Magnificent (*Fregata magnificens*) and Great (*Fregata minor*) (Fig. 3.14).

Some birds may have used ocean currents to arrive to the Galapagos, such as the Galapagos Penguin (*Spheniscus mendiculus*) and possibly also the ancestors of the Galapagos's now Flightlees Cormorant (*Phalacrocorax harrisi*; Snow and Nelson 1984; Valle 1986, 1993, 1994; Kennedy et al. 2009). The closest relative of the Galapagos penguin is the South American Humboldt penguin (*Spheniscus humboldti*), which is typical of more southern areas of South America. Smaller birds like the finches, doves, rails, or mocking birds probably arrived on drafts of vegetation carried by the ocean current, in particular the Humboldt Current (Fig. 3.15).

In the case of the famous Darwin's finches, genetic evidence indicates that the closest living relatives are the grassquits (Emberizidae) of Central and South America. They arrived to the Galapagos at least 1.5 million years ago, and are an excellent example of adaptive radiation, but their morphology is much more diverse than their genetic composition. Although there are low levels of movement between populations on the Galapagos, on some islands, natural selection has been strong enough to overcome the homogenization effects of gene flow and of hybridization. Divergence between species in isolation and the founder effect probably explain some of the difference, but

(a)
(b)

Fig. 3.15 **a** Widespread oceanic bird species: Blue Footed Booby, Isla Fernandina. *Photo* J. Bello-Page; **b** Endemic bird species: Galapagos Flightless Cormorant on nest, Punta Espinoza, Isla Fernandina. *Photo* K. N. Page

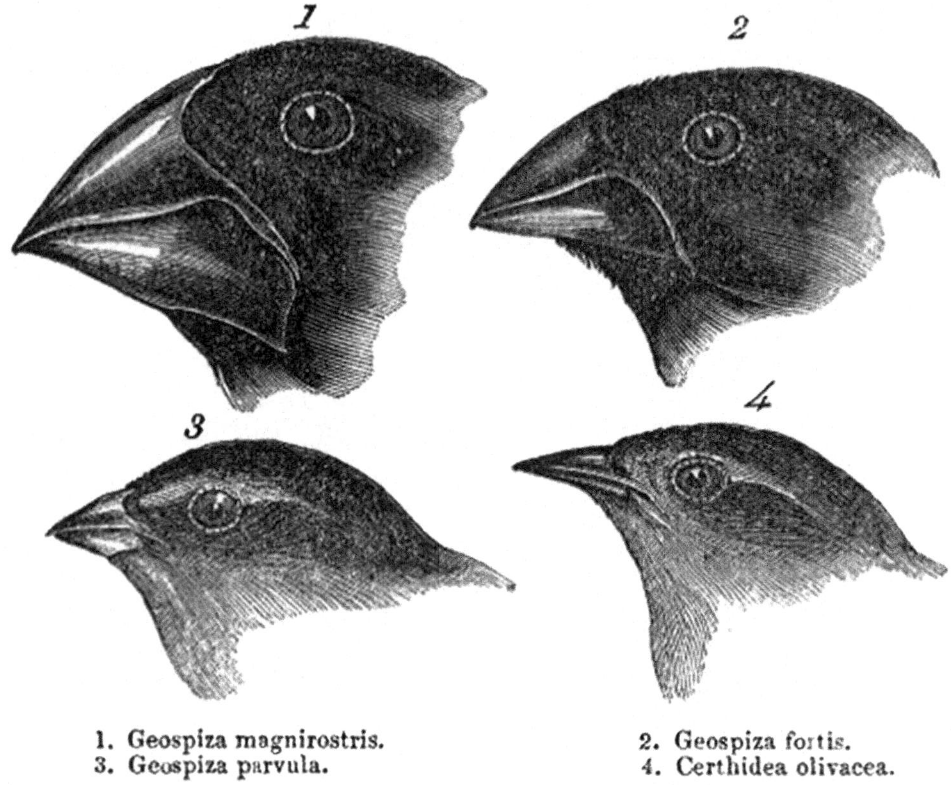

Fig. 3.16 Four examples of Darwin's Finches with beaks adapted to different diets (from the 1890 edition of Darwin's *Journal of Researches into the Geology and Natural History of the Various Countries Visited by H. M. S. Beagle*)

competition in sympatric conditions may also further differentiate species (Grant and Grant 2008) (Fig. 3.16).

The Galapagos mockingbirds, *Nesomimus* spp., are another example of adaptive radiation and inter-island speciation (Arbogast et al. 2006). The tropical mocking birds of northern South America (*Mimus. gilvus*) and the Caribbean mocking bird (*Mimus. gundlachii*) have been suggested as possible ancestors (Arbogast et al. 2006). New genetic studies, however, have indicated that the closest relatives are not the long tail mocking birds now found in Ecuador and Peru, but the North American *Mimus. polyglottos* author. Darwin (1845) realized that the mocking birds represented what we now know as allopatric evolution, and noticed that they were similar to the species he saw on the mainland—a key observation for the formulation of his theory of evolution (and more important than his superficial initial observations on the finches which he noted as being interesting but about which he could say little more: Sulloway 1982; Nichols 1997). Although there are only four species of mocking birds across the islands, in contrast to the finches, there is no case where two different species of mocking birds can be found in the same island: *N. melanotis* is found on San Cristobal, *N. macdonaldi* on Española, *N. triasciatus* on some islets around Floreana and a fourth, *N. parvulus*, distributed across several other islands (Arbogast et al. 2006).

The two species of sea lions present—the Galapagos Sea Lion (*Zalophus wollebecki*) and the Galapagos Fur Seal (*Arctocephalus galapagoensis*)—are endemic to the islands and probably arrived independently, some individuals even now occasionally making it back to the continental mainland. The Galapagos sea lion, is a sister species of the California sea lion (*Zolophus carlifornianus*) which is found only as far south as Baja California. In the Galapagos, sea lions seem to be on the edge of their environment, as they have to dive deeper to obtain food than those on the adjacent to the mainland (Páez-Rosas and Aurioles-Gamboa 2010; Villegas-Amtmann et al. 2008). According to the molecular clock, Californian and Galapagos sea lions had a common ancestor around 2.3 Ma (Wolf 2007). The other species of sea lion, the Galapagos Fur Seal, was named as such by the sealers of the nineteenth century, who greatly reduced their numbers. Both species live in colonies, with the Sea Lions typically being found on sandy beaches whilst the Fur Seals tend to live on the rocky shores. The colonies are controlled by a dominant male that defends the territory and the females from other male intruders, their larger size and greater bulk making then conspicuous when groups of Sea Lions are observed by visitors (Trillmich 1981) (Fig. 3.17).

There are two species of bats in the Galapagos, the hoary bat (*Lasiurus cinereus)* with three subspecies, and the eastern red bat (*Lasiurus borealis*). Both probably arrived from South or Central America, or possible even North America (Koopman and McCracken 1998). Of the seven species of rice rats (genera *Nesoryzomys* and Oryzomys) that were

Fig. 3.17 Sea-lions in their natural habitat, Isla Santiago. *Photo* J. Bello-Page

described some 150 years ago, only four were known to have survived to the year 2,000 (Dowler et al. 2000—although Clark 1984, considered that there were once 10 species…).

As many as seven lineages of reptile may have colonized the Galapagos, of which at least four diversified on the islands after their arrival. The best-known known examples of adaptation are the Galapagos tortoises. At least species have been identified which are believed to be part of a multispecies lineage. The closest living relative is the Chaco, a Chilean tortoise, *Geochelone chilensis* (Caccone et al. 1999), and the node of radiation in western southern Isabella is less than 500,000 years ago, and probably less for Fernandina (Caccone et al. 2002, 2004; Beheregaray et al. 2004).

During historical times, there were at least 15 species of Galapagos tortoise across the islands, although only 11 are now extant. Some authors consider them to be different subspecies (Pritchard 1996), whereas others refer to them be full species (Powell and Caccone 2006, 2008). Española, San Cristobal, Pinzon, Santiago and Pinta now have only one living species per island (it is believed by some that a second species on San Cristobal is now extinct). Santa Cruz, however, has two species and Isabela has one in each of the main volcanoes: Wolf, Darwin, Alcedo, Sierra Negra and Cerro Azul (Caccone et al. 2004; Parent et al. 2008) (Fig. 3.18).

The shape of the carapace of these reptiles gave the archipelago its name and provided some of the earliest evidence used by Charles Darwin to prove his theory of evolution. Tortoises have evolved saddle shape carapaces on low-lying islands that lack the water necessary to sustain lush vegetation; whereas those islands with large volcanoes have plenty of vegetation and have become the setting for the evolution of the dome-shaped forms of carapace. Tortoises are the largest herbivores on the Islands, the males reaching weights of 227 kg, with the females reaching up to 113 kg. Although no one knows with certainty, there are several theories about how the tortoises arrived on the islands. Some scientists, including Charles Darwin, believe they were carried by the currents in floating bodies of vegetation, probably some think during large El Niño events (Fig. 3.19).

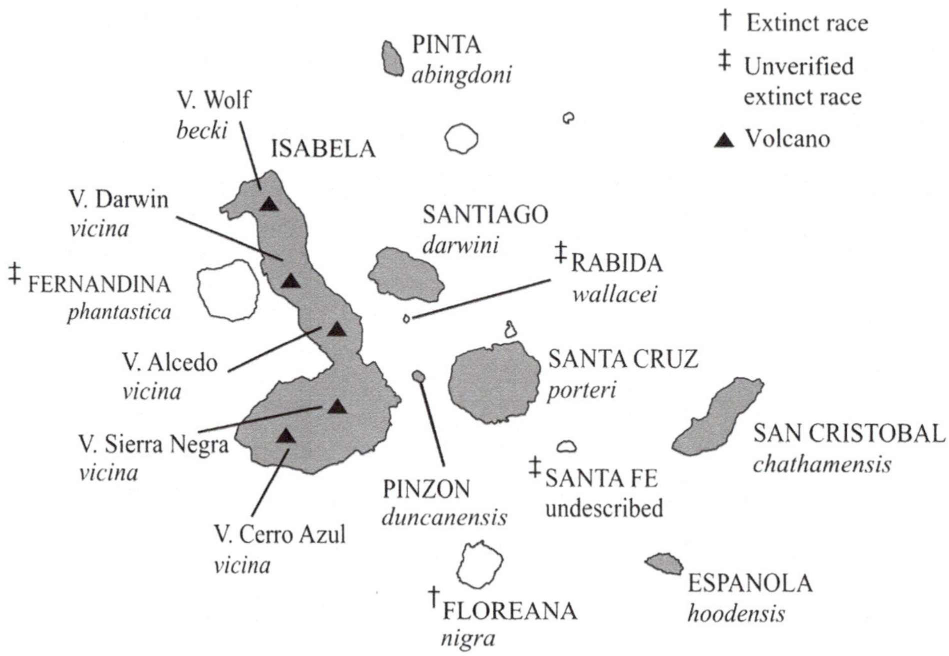

Fig. 3.18 The distribution of Galapagos tortoise species, ancient and modern (from https://commons.wikimedia.org/wiki/File:Galapagos_tortoise_distribution_Line_diagram.png)

Fig. 3.19 Giant Galapagos tortoises: in the wild on Isla Isabela (left) and in a tortoise ranch on Isla Santa Cruz (right). *Photos* K. N. Page/J. Bello-Page

Lava lizards (*Microlophus)* are also greatly diversified across the Galapagos, with seven species across the islands: *M. bivattatus*, *M. grayi*, *M. delanonis*, *M. habellii*, *M. pacificus*, *M. albemarlensis* and *M. duncanensis*,. The sister taxon is a species found in coastal Ecuador and Peru (probably *peruvianus*, *M. theresiae*, and *M. thoracicus*,) around 2 million years ago (Benavides et al. 2009). Genetic drift within this group apparently occurred during the Pleistocene when there was a greater connection between some of the islets and the Island of Santa Cruz, as has been determined using microsatellite markers (Jordan and Snell 2008) (Fig. 3.20).

Fig. 3.20 Lava lizard on Isla Isabela near Punto Moreno. *Photo* J. Bello-Page

The marine iguana *Amblyrrynchus* and the land iguana *Conolophus* are probably most closely related to the iguana *Ctenosaura* iguanas of Mexico and Central America (Miralles et al. 2017). The separation of these two closely related species is thought occurred on former Galapagos Islands, some 10–20 million years ago (Christie et al. 1992; Rassmann 1997). One of the most remarkable species of land iguana on the Galapagos is the famous 'Pink Iguana'—Conolophus marthae—which was only described in 2000 from Volcan Wolf in Isabella (Gentile 2009) (Fig. 3.21).

The best-known case of adaptation amongst modern reptiles is probably that of the marine iguanas *Amblyrrhychus cristatus.* Not only has its morphology evolved as it developed a flattened tail to aid swimming and diving, clawed fingers to grab onto the rocks, a short snout and modified teeth to enable grazing and a dark color to warm up fast on the rocks—it has also developed some physiological characteristics. These include the ability to respond to major changes in the availability of nutrients during strong El Niño events, when they decrease their body size and shorten their bones (Wikelski and Corinna 2000) but also a mechanism to secrete salt absorbed from sea water and their food. They have also a specialized internal microbiome that has evolved to allow them to digest the marine algae that they eat (Fig. 3.22).

The discovery of a series of seamounts representing the former islands opens a window of colonization some 17 million years ago. The "conveyer belt" mechanism was proposed by Axelrod (1972) as a general evolutionary scenario for many Pacific island biotas. Some authors have proposed it as a possible explanation for marine iguanas and other taxa (Wright 1983; Wyles and Sarich 1983; Rassmann 1997; Sequeira et al. 2008.

It is not clear if the Galapagos tortoises and the flightless Galapagos cormorants are good examples of the 'island rule' that says that small animals become larger and large animals become smaller, although in both cases these are larger than their mainland relatives. In many isolated islands there are also examples of dwarfism, such as the extinct miniature elephants (i.e. around 1 m high) recorded on some Mediterranean Islands (Poulakakis et al. 2002), and the small hominids of the island of Flores, in Indonesia (Losos and

Fig. 3.21 An endemic land iguana of the genus *Conolophus* on Isla Isabela. *Photo* K. N. Page

Fig. 3.22 The marine iguana, *Amblyrhynchus cristatus*—endemic to the Galapagos—on land and grazing marine algae for which its short snout is adapted (compare with Fig 3.21). *Photos* J. Bello-Page

Ricklefs 2009), although the Galapagos does not appear to have any clear examples of this phenomenon. However, the development of the relatively large and flightless Galapagos Cormorant (*Phalacrocorax harrisi*), maybe the best example of the 'island rule', as it represents an evolutionary pattern in many birds and insects to become flightless, due a lack of typical terrestrial predators, which may also be linked to an overall increase in body size. The loss of dispersal abilities also found in plants was considered by Darwin to be an adaptation to oceanic islands, so that the risk of long distance dispersal and hence loss is minimized (Losos and Ricklefs 2009; Whittaker and Fernandez Palacios 2007).

3.4 Marine Fauna

Marine animals, especially large pelagic fish are migratory, moving in the oceans and they never evolving into endemic species. In contrast, reef fish, and benthic and sessile invertebrates, many of which arrived to the islands as larva, have a relatively high proportion of endemic species. Changes in oceanic currents, between 4.5 and 3.6 Ma, caused by the closure of the Isthmus of Panama (Haug and Tiedemann 1998) contributed to the development of the Galapagos distinct marine biota. This closure generated a major alternation in the pattern of dispersal of many marine organisms around the time when the first of the existing islands were emerging. Of the Galapagos marine organisms 11.4% of the fish, 26% of the Polychaeta worms, 21.5% of the porcelain crabs, 22.2% of the barnacles, 18.1% of the mollusks, 17% of the echinoderms, 30% of the algae, and 39.5% of the ahermatypic corals are endemic (Bustamante et al. 2002). In the case of some crustaceans and mollusks, however, there is possible connectivity between the mainland and the Galapagos, meaning that occasional genetic flow tends to homogenize the organisms of Galapagos with those of the mainland.

Some of the marine invertebrates, such as the porcelain crabs (Porcellanidae), have short dispersal larva phase which makes it more difficult for them to disperse from distant coastal areas; of the 92 species found in the Pacific only 14 species are found in the Galapagos (Harvey 1991). Most mollusks, however, have larva that can cross large distances, which is reflected in the Indo Pacific affinities of 13 of the species of Galapagos mollusks. Due to temperature-related adaptations, most of these mollusks come from the Panamic-Californian region (Finet 1991).

One of the most conspicuous marine inhabitants, however, are the Galapagos Green Turtles, *Chelonia agassizii*. There is an open debate on their affinities, with some considering them to be a subspecies or simply a local population of the Green Sea Turtle, *Chelonia mydas*, whilst others consider them to be a species in their own right (see https://en.wikipedia.org/wiki/Gal%C3%A1pagos_green_turtle which has a useful set of primary sources listed). Whatever the reality, the species would have arrived in the Galapagos by actively swimming, but the distance from other nesting populations, for instance on the continental mainland, would inevitably mean that genetic exchange would be limited and the development of a local population with some distinct features could occur (Fig. 3.23).

3.5 Invasive Species

Oceanic islands are places where specific adaptations have made some of the endemic and native plants and animals more vulnerable to exotic species (Krajick 2005). This vulnerability is often a result of a lack of competitors or predators which has made them less able to compete or defend themselves. Introduced grazing animals such as goats and predators such as cats and dogs can, therefore, become serious threats.

Colorful introduced plants may attract pollinators more readily than the duller native plants, leading to the disadvantaging of the latter. A good example of this process problem concerns plants of the genus *Lantana*. Native Lantana, (*Lantana peduncularis*) cannot compete with its more colorful introduced cousin, *Lantana camara*, as some of the introduced insect species will be more likely to pollinate the introduced plant, thus creating a synergetic relationship between introduced plant and animal species, hence promoting the spread of the intruded plant, rather than the native species. Such relationships can also develop between introduced plants and some of the native species, such as tortoises and birds, that then become a vector helping g spread the seeds of the invasive plants (Quiroga and Rivas 2016).

An example of some of strategies which can make invasive species so successful, is the invasive hill raspberry (*Rubus niveus*), originally from Southern Asia, which has multiple reproductive strategies, including a capacity to produce thousands of seeds that can remain viable for long periods of time, but also to spread vegetative. Other invasive plants such as guava, trees like cedar tree (*Cedrela odorata*), breadfruit tree (*Artocarpus altilis*) and quinine tree (*Cinchona pubescens*) can have similar strategies and it is not surprising, therefore, that they can easily displace native and endemic species.

Invasive insects are probably one of the more difficult groups of animals to control, let alone eliminate, and include several species of ants, wasps and the parasitic fly, *Philornis downsi,* which threatens some of the most sensitive populations of birds, such as the mangrove finch (*Geospiza heliobates*) and the Floreana Mocking bird (*Neomimus trifasciatus*) (Quiroga and Rivas 2016).

Fig. 3.23 The Galapagos Green Turtle is common around the islands and is the only turtle to nest on the islands—its relationship to other Pacific populations of Green Sea Turtle is under debate. *Photo* J. Bello-Page

The aggressive process of displacement of some of the main native species has generated different responses from conservation agencies and scientists (Hobbs et al. 2009, 2013; Murcia et al. 2014), although typically involving complete elimination all the invasive species, or the introduction of biological controls. As some of these eradication efforts, such as with *Rubus niveus,* have failed, some scientists are beginning to argue that it is necessary to create conservation strategies that accept some of the invasive plants and animals as part of the ecosystems (Rentería et al. 2012). This idea of valuing, or at least tolerating the new ecosystems being created is refers to 'novel' and 'hybrid' ecosystems (Hobbs et al. 2009, 2013), and has caused a considerable controversy among conservationists. One of the problems with invasive species is that they create symbiotic relations with native and endemic species which means that the elimination of some of these invasive plants will affect the population of some of the native species of the Galapagos. Such could be the result of the elimination of some of the introduced trees that now sustain populations of endemic epiphytes. In addition, despite high profile initiatives to eliminate introduced grazing animals such as goats (Campbell et al. 2004; Cruz et al. 2009), their presence locally in small and controlled numbers may actually help prevent the colonization and loss of some more open habitats by invasive scrub (in the absence of sufficient numbers of endemic grazers (i.e. tortoises).

Hybridization between endemic plants and animals and the recent arrivals is however, a more difficult problem and could be possible in the case of some plants. This is occurring between the native guava (*Psidium galapageium*) with the introduced species (*Psidium guayaba*), as well between for the two native species (*Solanum cheesmaniae* and *S. galapagense)* and the two introduced species of tomatoe (*S. lycopersicum* and *S. pimpinellifolium*), and between native (and introduced Lantanas (*Lantana peduncularis* and *Lantana camara* respectively). Hybridization could also become a problem for some animals such as the eight endemic species of Galapagos Gecko (*Phyllodactylus galapagoensis, P. bauriingtonensis, P. gorii, P. darwini, P. baurii, P. leei, P duncanensis* and *P. giberti*) with the five introduced species (*Phyllodactylus reissii, Phyllodactylus tuberculosus, Hemidactylus frenatus, Gonatodes caudiscutatus*, and *Lepidodactylus lugubris*) (Quiroga and Rivas 2016).

3.6 Conclusion

The Galapagos, as is the case with many oceanic islands, has developed a unique set of plants and animals and their evolution is closely related to the geology and geography of the Islands. Their geological history, their complexity as an archipelago, and their distance from the other land masses and islands has produced a unique set of characteristics that impressed Charles Darwin when he arrived in 1835. His ideas have been used by scientists ever since to understand the evolution of animal and plant species, as one of the most important discoveries of modern times. As is the case of other oceanic islands, the Galapagos are the backdrop against which impressive evolutionary solutions have been tried. Some of these forms have been successful despite their uniqueness, such as with the flightless cormorants, the vampire finches and the marine iguana. Others are of interest to scientists because of what they teach us about evolutionary process, such as adaptive radiation, including the daisy-tree *Scalesia*, the Darwin finches, the tortoises, the lava lizards and the land snails. In the Galapagos, however, it is the geological characteristics of the islands that have created the backdrop for the evolution of these species and their diversification. Key factors include the relative ages of the islands, and their predecessors, their distance from the mainland, the distance between islands and their elevation—including the potential to produce relatively fertile soils over time under the moister conditions associated with altitude (and hence contributing to the development of diverse niches for colonization). The volcanic origins and emergence of each island has created a succession of empty niches into which the few species that arrived could adapt. This adaptation meant that a series of phases of diversification were generated by a process based on the geological and geographical characteristics and evolution of the islands—it is this relationship between geology, ecology and evolution that underlies the importance of the Galapagos as a 'natural laboratory' for evolutionary studies.

The very same characteristics that make the animals so unique and special for scientists, conservationists and tourists, such as their tameness, their small population size, the degree of endemism, also makes them vulnerable to introduced species. As the connectivity between the islands and the mainland increases and as the number of tourists grows and as the islands' economy improves, there will be more introduced species coming to the islands. This constant trend will include not only the arrival of aggressive and invasive plant, insect and animal species, but also the introduction of diseases and their vectors. Some of these diseases could affect the survival of some of the most charismatic animals such as penguins, flamingos and cormorants. This threat could be compounded as climate change, contaminating spills and land conversion increases the vulnerability of the wildlife populations. Thus, the challenges facing not only the Galapagos but many other oceanic islands are great—i.e. the very same characteristics that make oceanic islands so interesting from an evolutionary perspective, and have made it possible for many of its animals and plants to evolve, are also responsible for the fragility of their native animals and plants.

References

Arbogast BS, Drovetski SV, Curry RL, Boag PT, Seutin G, Grant PR, Grant BR, Anderson DJ (2006) The origin and diversification of Galapagos mockingbird. Evolution 60(2):370

Axelrod DI (1972) Ocean floor spreading in relation to ecosystematic problems. In: Allen RR, James FC (eds) Symposium on ecosystematics. University of Arkansas Press, Fayetteville, AR, pp 15–76

Barthlott W, Porembski S (1994) Die Kakteen der Galápagos-Inseln. In: Zizka G & Klemmer K (Hrsg) Pflanzen- und Tierwelt der GalápagosInseln.—Kleine Senckenbergreihe 20, Palmengarten Sonderheft 22: 76–80

Beheregaray LB et al (2004) Giant tortoises are not so slow: rapid diversification and biogeographic consensus in the Galapagos. Proc National Acad Sci 101:17. https://doi.org/10.1073/pnas.0400393101

Benavides E, Baum R, Snell HM, Snell HL, Sites W (2009) Island biogeography of Galapagos lava lizards (Tropiduridae: *Microlophus*): species diversity and colonization of the Archipelago the society for the study of evolution. Evolution 63–6:1606–1626

Breure ASH (1979) Systematics, phylogeny and zoogeography of Bulimulinae (Mollusca). Zoologische Verhandelingen, Leiden, Germany

Bustamante RH, Wellington GM, Branch GM, Edgar GJ, Martinez P, Rivera F, Smith F, Witman J (2002) Outstanding Marine Features. In: Bensted-Smith R (ed) A biodiversity vision for the Galapagos Islands. Charles Darwin Foundation and World Wildlife Fund, Puerto Ayora, pp 60–71

Caccone A, Gibbs JP, Ketmaier V, Suatoni E, Powell JR (1999) Origin and evolutionary relationships of giant Galapagos tortoises. Proc Natl Acad Sci USA 96:13223–13228. https://doi.org/10.1073/pnas.96.23.13223

Caccone A, Gentile G, Gibbs JP, Fritts TH, Snell HL, Betts J, Powell JR (2002) Phylogeography and history of giant Galapagos tortoises. Evolution 56:2052–2066. https://doi.org/10.1111/j.0014-3820.2002.tb00131.x

Caccone A, Gentile G, Burns C, Sezzi E, Bergman W, Powell JR (2004) Extreme difference in rate of mitochondrial and nuclear

DNA evolution in a large ectotherm. Galapagos tortoises. Mol. Phylogenet, Evol, p 31
Campbell KJ, Donlan CJ, Cruz F, Carrion V (2004) Eradication of feral goats (Capra hircus) from Pinta Island, Galapagos, Ecuador. Oryx 38:1–6
Chambers SM (1991) Biogeography of Galapagos land snails. In: James MJ (ed) Galapagos marine invertebrates. Plenum, New York, pp 307–325
Cruz F, Carrion V, Campbell KJ, Lavoie C, Donlan CJ (2009) Bio-economics of large-scale eradication of feral goats from Santiago Island, Galapagos. J Wildl Manag 73(2):191–200
Christie DM, Duncan RA, Mc Birney AR, Richards MA, White WM, Harpp KS, Fox CG (1992) Drowned islands downstream from the Galapagos hotspot imply extended speciation times. Nature 355:246–248
Clark DA (1984) In: Perry R (ed) Native land mammals in key environments: Galapagos. Pergamon Press, Oxford, UK, pp 225–231
Darwin CR (1845) Journal of researches into the natural history and geology of the countries visited during the voyage of H.M.S. Beagle round the world, under the Command of Capt. FitzRoy RN, John Murray, London, UK
Dowler R, Carroll DS, Edwards CW (2000) Rediscovery of rodents (Genus Nesoryzomys) considered extinct in the Galapagos Islands. Oryx 34(2):109
Finet Y (1991) The marine molluscs of the Galápagos Islands. In: James MJ (ed) Galápagos Marine Invertebrates: Taxonomy, Biogeography, and Evolution in Darwin's Islands. Plenum, New York
Fleischer RC, McIntosh CE, Tarr CL (1998) Evolution on a volcanic conveyor belt: using phylogeographic reconstructions and K–Ar-based ages of the Hawaiian Islands to estimate molecular evolutionary rates. Mol Ecol 7:533. https://doi.org/10.1046/j.1365-294x.1998.00364.x
Gentile G (2009) Iguanas terrestres: Descubrimento de una nueva especie. In: de Roy T (ed) Galapagos: cincuenta anos de ciencia y conservacion. Parque Nacional Galapagos, pp 114–121
Grant PR, Grant BR (2008) How and why species multiply: the radiation of Darwin's finches. Princeton University Press, Princeton, NJ
Harvey AW (1991) Biogeographic patterns of the Galápagos porcelain crab fauna. In: James MJ (ed) Galápagos marine invertebrates: taxonomy, biogeography, and evolution in Darwin's Islands. Plenum, New York, pp 157–172
Haug GH, Tiedemann R (1998) Effect of the formation of the Isthmus of Panama on Atlantic Ocean thermohaline circulation. Nature 393:673–676
Hobbs R, Higgs E, Harris J (2009) Novel ecosystems implications for conservation and restoration. Trends Conserv Evol 24(11):599
Hobbs RJ, Higgs ES, Hall CA (eds) (2013) Novel ecosystems: intervening in the new ecological world order. Wiley-Blackwell, Oxford, UK
Ingala P, Orstom I (1989) Inventario Cartográfico de los Recursos Naturales, Geomorfología, Vegetación, Hídricos, Ecológicos y Biofísicos de las Islas Galápagos, Ecuador. 1:100,000 maps. INGALA, Quito
Jordan MA, Snell HL (2008) Historical fragmentation of islands and genetic drift in populations of Galapagos lava lizards (Microlophus albermarlensis complex). Mol Ecol 14:859–867. https://doi.org/10.1111/j.1365-294x.2005.02452.x
Juan C, Emerson BC, Oromí P, Hewitt GM (2000) Colonization and diversification: towards a phylogeographic synthesis for the Canary Islands. Trends Ecol Evol 15(3):104–109
Koopman KF, McCracken GF (1998) The taxonomic status of Lasiurus (Chiroptera: Vespertilionidae) in the Galapagos Islands. American Museum of Natural History
Kennedy M et al (2009) The Phylogenetic position of the Galápagos cormorant. Mol Phylogenetics Evol 53(1):94. https://doi.org/10.1016/j.ympev.2009.06.002
Kizirian D, Trager A, Donnelly MA, Wright JW (2004) Evolution of Galapagos Island lava lizards (Iguania: Tropiduridae: Microlophus). Mol Phylogenet Evol 32:761–769. https://doi.org/10.1016/j.ympev.2004.04.004
Kohn AJ (1972) Conus-miliaris at Easter Island—ecological release of diet and habitat in an isolated population. Am Zool 12:712
Krajick K (2005) Winning the war against island invaders. Science 310:1412. www.sciencemag.org
Landry B (2002) Galagete, a new genus of Autostichidae representing the first case of an extensive radiation of endemic Lepidoptera in the Galapagos Islands. Rev Suisse Zool 109:813–868
Losos JB, Ricklefs RE (2009) Adaptation and diversification on islands. Nature 457:12
Macarthur RH, Wilson EO (1967) The theory of Island biogeography. REV—Revised ed., JSTOR, Princeton University Press. www.jstor.org/stable/j.ctt19cc1t2
McMullen CK (1990) Reproductive biology of Galapagos Islands angiosperms. In: Lawesson JA, and others ed(s). Botanical research and management in Galapagos. Proceedings of a workshop on… held 11–18 April 1987 at the Charles Darwin Research Station, Santa Cruz, Galapagos, Ecuador. St. Louis, Mo., Missouri Botanical Garden (Monographs in Systematic Botany), vol 32, pp 35–45—illus., map.. En Reproductive biology. Geog Floristics (SOUTH_-AMERICA: GALAPAGOS_ISLANDS)
McMullen CK (1987) Breeding systems of selected Galápagos Islands angiosperms. Am J Bot 74:1694
McMullen CK (1989a) The Galápagos carpenter bee, just how important is it? Noticias de Galápa-gos 48:16
McMullen CK (1989b) Flowering colonizers of the Galápagos Islands: drab but not dull. *Plants Today* 2
McMullen CK (1990) Reproductive biology of Galápagos Islands angiosperms. Monographs in Systematic Botany from the Missouri Botanical Garden 32
McMullen CK (1999) Flowering plants of the Galapagos. Cornell University Press
Miralles A, Macleod A, Rodríguez A, Ibáñez A, Jiménez-Uzcategui G, Quezada G, Vences M, Steinfartz S (2017) Shedding light on the imps of darkness: an integrative taxonomic revision of the Galápagos Marine Iguanas (Genus Amblyrhynchus). Zool J Linn Soc 10:1–33
Murcia C, Aronson J, Kattan GH, Moreno-Mateos D, Kingsley D, Simberloff D (2014) A critique of the "novel ecosystem" concept. Trends Ecol Evol 29:548–553
Paulay G (1994) Biodiversity on Oceanic Islands: its origin and extinction. Integr Compar Biol 34(11):134–144. https://doi.org/10.1093/icb/34.1.134
Parent CE, Caccone A, Petren K (2008) Colonization and diversification of Galapagos terrestrial fauna: a phylogenetic and biogeographical synthesis. Philosophical Transactions of the Royal Society B: Biological Sciences
Parent CE, Crespi BJ (2006) Sequential colonization and diversification of Galapagos endemic land snail genus Bulimulus (Gastropoda, Stylommatophora). Evolution 60:2311–2328
Parent C, Crespi BJ (2009) Source ecological opportunity in adaptive radiation of Galápagos endemic land snails. Am Nat 174(6):898–900
Páez-Rosas D, Aurioles-Gamboa D (2010) Alimentary niche partitioning in the Galapagos Sea Lion Zalophus Wollebaeki. Marine Biol 157(12):2769–2781. https://doi.org/10.1007/s00227-010-1535-0
Peck SB (2006) The beetles of the Galapagos Islands, Ecuador: evolution, ecology, and diversity (Insecta: Coleoptera). NRC Research Press, Ottawa, Canada

Porter DM (1983) Vascular plants of the Galapagos: origins and dispersal. In: Bowman RI, Berson M, Leviton AE (eds) Patterns of evolution in Galapagos organisms. San Francisco, California, Pacific Division of AAAS, pp 33–96

Poulakakis N, Mylonas M, Lymberakis P, Fassoulas C (2002) Origin and taxonomy of the fossil elephants of the island of Crete (Greece): problems and perspectives. Palaeogeogr Palaeoclimatol Palaeoecol 186(1–2):163–183. https://doi.org/10.1016/s0031-0182(02)00451-0

Powell JR, Caccone A (2006) A quick guide to Galapagos tortoises. Curr Biol 16:R144–R145. https://doi.org/10.1016/j.cub.2006.02.050

Powell JR, Caccone A (2008) CSI tortoise: unraveling the mystery of mysteries. Galapagos News 26:8–9

Pritchard PCH (1996) The Galapagos tortoises: nomenclatural and survival status Lunenburg. Chelonian Res Found, MA

Quiroga D, Rivas G (2016) Restoration: novel ecosystems and hybrid environments. In: Quiroga D, Sevilla A (eds) Darwin, Darwinism and conservation. Springer, NY

Raia P, Meiri S (2006) The Island rule in large mammals: paleontology meets ecology. Evolution 60(8):1731–1742. https://doi.org/10.1111/j.0014-3820.2006.tb00516.x

Rassmann K (1997) Evolutionary age of the Galapagos iguanas predates the age of the present Galapagos Islands. Mol Phylogen Evol 7(2):158–172

Rentería JL, Gardener MR, Panetta FD, Atkinson R, Crawley MJ (2012) Possible impacts of the invasive plant Rubus niveus on the native vegetation of the Scalesia forest in the Galapagos Islands. PLoS ONE 7(10):e48106. https://doi.org/10.1371/journal.pone.0048106

Schmitz P, Cibois A, Landry B (2007) Molecular phylogeny and dating of an insular endemic moth radiation inferred from mitochondrial and nuclear genes: the genus Galagete (Lepidoptera: Austostichidae) of the Gala´pagos Islands. Mol Phylogenet Evol 45. https://doi.org/10.1016/j.ympev.2007.05.010

Sequeira AS, Lanteri AA, Albelo LR, Bhattacharya S, Sijapati M (2008) Colonization history, ecological shifts and diversification in the evolution of endemic Galápagos weevils. Mol Ecol 17(4):1089–1107

Snow DW, Nelson JB (1984) Evolution and adaptations of Galapagos sea-birds. Biol J Linn Soc 21(1–2):137–155. https://doi.org/10.1111/j.1095-8312.1984.tb02057.x

Stocklin J (2009) Darwin and the plants of the Galápagos-Islands. Bauhinia 21:33–34

Sulloway FJ (1982) Darwin and his finches: the evolution of a legend. J Hist Biol 15(1):1–53. https://doi.org/10.1007/bf00132004

Traveset A, Heleno R, Chamorro S, Vargas P, McMullen CK, Castro-Urgal R, Nogales M, Herrera HW, Olesen JM (2013) Invaders of pollination networks in the Galapagos Islands: emergence of novel communities. Proc Royal Soc B: Biol Sci 280 (1758):20123040. https://doi.org/10.1098/rspb.2012.3040

Trillmich F (1981) Mutual mother-pup recognition in Galápagos fur seals and sea lions: cues used and functional significance. Behaviour 78(1/2):21–42

Trueman M, d'Ozouville N (2010) Characterizing the Galapagos terrestrial climate in the face of global climate change by: Galapagos Research 67:26–37

Tye A, Ortega JF (2011) Origins and evolution of Galapagos endemic vascular plants. The biology of Island floras. Cambridge University Press, Cambridge

Valle C (1986) Status of the Galápagos penguin and flightless cormorant. Noticias Galápagos 43:16–17

Valle CA (1993) The evolution of mate desertion in flightless cormorants. In: Merlen-Davis G (ed) Charles Darwin Research Station Annual Report 1988–1989. Charles Darwin Foundation, Santa Cruz, Galapagos, Ecuador, pp 97–99

Valle CA (1994) The ecology and evolution of sequential polyandry in Galápagos cormorants (Compsohalieus [Nannopterum] harrisi). Ph. D. Dissertation, Princeton University, Princeton, USA

Valle CA, Coulter MC (1987) Present status of the flightless cormorant, Galápagos penguin and greater flamingo populations in the Galápagos Islands, Ecuador, after the 1982–1983 El Niño. Condor 89:276–281

Van Den Bergh GD, Rokhus DA, Morwood MJ, Sutikna T, Jatmiko, Saptomo EW (2008) The youngest *Stegodon* remains in Southeast Asia from the Late Pleistocene archaeological site Liang Bua, Flores, Indonesia. Q Int 182(1): 16–48. https://doi.org/10.1016/j.quaint.2007.02.001. Retrieved 27 Nov 2011

Vargas P, Heleno R, Traveset A, Nogales M (2012) Colonization of the Galapagos Islands by plants with no specific syndromes for long-distance dispersal: a new perspective. Ecography 35(33–43):2012

Villegas-Amtmann S, Costa DP, Tremblay Y, Salazar S, Aurioles-Gamboa D (2008) Multiple foraging strategies in a marine apex predator, the Galapagos sea lion Zalophus wollebaeki. Mar Ecol Prog Ser 363:299–309

Walters SM, Stow EA (2002) Darwin's mentor. Cambridge University Press. ISBN 0-521-59146-5

Wagner WL, Funk VA (eds) (1995) Hawaiian bio- geography: evolution on a hot spot archipelago. Smithsonian Institution Press, Washington, DC

Walsh SJ, McCleary AL, Mena CF, Shao Y, Tuttle JP, González A, Atkinson R (2008) QuickBird and Hyperion data analysis of an invasive plant species in the Galapagos Islands of Ecuador: implications for control and land use management. Remote Sens Environ 112 (5):1927–1941. https://doi.org/10.1016/j.rse.2007.06.028

Weston EM, Lister AM (2009) Insular dwarfism in hippos and a model for brain size reduction in Homo floresiensis. Nature 459(7243):85

Whittaker R, Fernandez P (2007) Island biogeography, ecology, evolution and conservation. Oxford University Press, Oxford

Wiggins IL, Porter DM (2018) Flora of the Galapagos Islands. Stanford University Press, 1971, p 1020. www.sup.org/books/title/?id=3. Accessed 8 Apr 2018

Wikelski M, Corinna T (2000) Marine iguanas shrink to survive El Niño. Nature 403

Wolf M et al (2007) The evolution of animal personalities. Animal Personalities 8 Nov. 2007, pp 252–275. https://doi.org/10.7208/chicago/9780226922065.003.0010

Wright JW (1983) The evolution and biogeography of the lizards of the Galapagos Archipelago: evolutionary genetics of *Phyllodactylus* and *Tropidurus* populations. In: Bowman RI, Berson M, Leviton AE (eds) Patterns of evolution in Galápagos organisms. Pacific Division of the American Association for the Advancement of Science, San Francisco, CA, pp 123–155

Wyles JS, Sarich VM (1983) Are the Galápagos iguanas older than the Gal´apagos. In: Bowman RI, Berson M, Levinton AE (eds) Patterns of evolution in Galapagos organisms. American Association for the Advancement of Science, Pacific Division, San Francisco, CA, pp 177–185

4 Geoconservation, Geotourism and Sustainable Development in the Galapagos

4.1 Science and Nature Conservation: Origins of a Concept

4.1.1 Nature: From Romantic Associations to Collection

The concept of nature experienced an important shift from the 19th into the early 20th century as people started to conceive conservation as a scientific and even professional endeavor. The tradition of finding truth by observing nature and developing explanations of its workings, which had existed in Western European cultures since the ancient Greek philosopher Aristotle's time (4th Century BC,) became later an important aspect of modernity. However, modern views have become increasingly mechanistic in their understanding, leading to a disenchantment with nature (Botkin 2012).

In the 18th and earlier 19th centuries 'conservation' of fauna and flora was essentially no more than an effort to collect and preserve the specimens in the political and academic centers of the time. This was very much the context of Darwin's role when he visited the islands, especially as a large part of the animals and plants that he collected were endemic to the islands—a fact that would later prove to be critical in the development of the theory the evolution of the species. Perhaps one of the greatest stimuli to this campaign of documentation was the establishment of a formal structure and methodology for naming and classifying all living (and fossil) species by the Swedish naturalist, Carl Linneous in the 18th Century (https://en.wikipedia.org/wiki/Carl_Linnaeus). Using latinised binomia to group named species into genera, and thence into higher 'orders' he was also perhaps one of the first to systematically demonstrate relationships between different animals and plants, something which Darwin would argue was due to evolutionary processes.

From the later 17th century, naturalists and collectors following in the footsteps of Linneous started travelling to many regions of the world to bring back to Europe large numbers of specimens and the most prestigious scientific institutions amassed a large number of specimens in the European centers of power. These efforts were sponsored by emerging intuitions like the British and the French Academies of Science. Indeed the vary naming of species could have 'imperial ramifications' for when a species is "*... named and described it becomes ... a possession forever and the value of every included specimen of it, even in a mercantile view, is enhanced*" and science becomes a type of "*metaphoric appropriation*" (Kirby 1825 as quoted in Desmond and Moore 1991, p. 343).

The idea of collecting to 'preserve' species was transformed with the development of a concept of ecological conservation and the idea of conserving plants and animals in their natural habitat. By the end of the 19th century, conservation had become an activity that also required the use of a scientific method (Bonham-Carter 1971). Together with the process of urbanization, technological improvement and rapid industrialization in many developed countries in 18th and 19th centuries there was an accelerated process of destruction of natural habitats. During this time of transformation, the romantic and the rational view of nature became the two contrasting and to some extent complementary dominant views of nature in the western European-influenced cultures. Thus from an historic perspective, the idea of conservation of nature gained strength as industrial development and agricultural intensification displaces natural areas.

Part of the cultural change that was generated during this time was the idealization of natural landscapes. By the mid-eighteenth century, '*something quite new in history was making its presence known: the collective enjoyment of the scenery of nature for its own sake and in its original, unmodified condition*' (Allen 1976). The appreciation of the beauty of nature became a main motivation for the protection of areas. A view that was originally promoted by the German naturalist and geologist, Alexander von Humboldt (1769–1859; https://en.wikipedia.org/wiki/Alexander_von_Humboldt) and painters like the American Edwin Church (1826–1900; https://en.wikipedia.org/wiki/Frederic_Edwin_Church) and

D. Kelley et al., *In the Footsteps of Darwin: Geoheritage, Geotourism and Conservation in the Galapagos Islands*, Geoheritage, Geoparks and Geotourism, https://doi.org/10.1007/978-3-030-05915-6_4

other intellectuals became a standard among the conservationists. Nature was now seen as indivisible by these romantic thinkers (Vesey-Fitzgerald 1969), a system that should be protected. Whereas taming nature was the goal of much of the early modern period, as represented by formal aristocratic estates and gardens in France and in other European countries, untamed nature was later also seen to be desirable. Thus although for hundreds of years, humans had been trying to establish a control over nature and turning wild nature into a garden, with the spread of industrialization, the prospect of a complete removal or systematic 'control' of nature, led untouched nature became more valued and interesting.

In the modern times, starting in the late 18th Century, museums, as well as scientifically-organized zoological and botanic gardens, became key facets of the effort to collect and *preserve* nature from remote areas. The major western countries, museums and private collectors around the world started to compete to create large collections, that were used by the growing group of scientists (both affluent 'amateurs' and the first institutional employed professionals) to reach conclusions about the origin, distribution and transformation of species. Such institutions also became a sign of national and private power and such activities a way to increase the both the status of the institution and the countries that sponsored them. It was seen also as an effort to *conserve* the animals and plants for the scientific study of the organisms. Imperial powers also collected—although 'plundered' might be a more appropriate term in some contexts—cultural and natural artifacts and specimens, including from within Europe itself—classical archaeology from Greece and other areas being especially targeted. The most infamous example of the latter are perhaps the so-called 'Elgin Marbles' from the ancient Greek Parthenon in Athens, 'purchased' by a British Aristocrat from the Turkish authorities then occupying the city and sold to the national British Museum, where they still reside, despite ongoing demands for their return (https://en.wikipedia.org/wiki/Elgin_Marbles). These objects and animals became a symbol to be displayed by empires, hence demonstrating their power and dominion over large areas of the world (Fig. 4.1).

In the 19th Century elements of the same elites started to realize that they would also to have to protect the vanishing nature around them, *including* in their own countries. In North America and especially western and central Europe, overhunting and habitat change had significantly affected local animal and plant populations, with massive decreases in numbers, even local extinctions, being noted. Organizations such as the Royal Society for the Protection of Birds in the UK, with its origins going back to 1889 (https://www.rspb.org.uk/about-the-rspb/about-us/our-history/) due to concern about wholesale and the virtually unregulated killing of birds for sport and decoration, and not just for food. In Europe, such clubs and societies started to mobilize people to understand and protect nature, later these groups would become important conservation societies. The protection of birds thus became an important pillar of the conservation movement (Nicholson 1970) and is still one of its strongest campaigning movements today. The Zoological Society of London, formed in 1826 (https://www.zsl.org/about-us), was one of the earliest scientifically focused organizations concerned with both captive and wild animals. It established London Zoo in 1828, an institution which is now a world leader in the conservation of endangered species, and inspired many other organizations to become established and follow similar principles.

A slightly different perspective on the relationship between human society and nature is represented by the slightly earlier founding of the Royal Society for the Prevention of Cruelty to Animals in 1824 (https://www.rspca.org.uk/whatwedo/whoweare/history) also in the UK, and although it was mostly concerned with domestic animal protection it had an important influence in the support for the British, 1869 Sea Birds Preservation Act.

Crucially, it was not only animals that became the focus of early protection measures; plants were also a matter of concern at the time. Botanical societies were created in many European and North American countries. In France, the Jardin des Plantes in Paris, created in the early 18th century, was expanded by the Compte de Buffon in 1739 and became a major resource for scientists with collections from around the world (http://www.jardindesplantes.net/). In the UK, the Botanical Society of London was created in 1836 (https://en.wikipedia.org/wiki/Royal_Botanic_Society) and also supported research, including the production of atlases of plant species, even and becomming involved in conservation efforts. This activity was strongly linked to the formation of the Linnean Society of London in 1788, which had acquired, and still houses, much of Linnaeus original systematic reference material, having purchased it from his Swedish heirs (https://www.linnean.org/the-society). In the USA, the foundation of the Audubon Society in 1896, represents an expression of the realization of the same concerns about the destruction of natural habitats and species in a 'new' and developing country (https://www.audubon.org/about/history-audubon-and-waterbird-conservation). The naming of the Society celebrated the inspirational wildlife artist, John James Audubon (1785–1851), whose monographic work on the Birds of America was completed in 1838—although he had had to sail to Europe to find a printer for his work (https://www.audubon.org/content/john-james-audubon).

With the development of intense debates about the origin of life, catalyzed by pioneers such as the French anatomists Jean-Baptiste Lamarck (1744–1829; https://en.wikipedia.org/wiki/Jean-Baptiste_Lamarck) and Georges Cuvier (1769–1832; https://en.wikipedia.org/wiki/Jean-Baptiste_Lamarck),

Fig. 4.1 A typical 19th century colonial museum, the former Hunterian Museum in London, crammed to the ceiling with exhibits—including skeletons of two giant South American mammals in the foreground (a giant ground sloth to the left and a giant anteater to the right) (As published in the *Illustrated London News* in 1844)

and Darwin himself, and as the theories of biological evolution became more popular but also more controversial in the 19th century, museum collections became essential for settling the academic disputes about the origin of life and the relevance—and even existence—of a God (Desmond and Moore 1991). As scientists started traveling more, it became clear that many of these disputes had to be settled by not only by using existing collections but also by looking at natural processes, in situ.

4.1.2 The Evolution of a Scientific View of Nature and the Establishment of Protected Areas

At the beginning of the 20th century various scientific ideas had become established a part of a 'Western' (i.e. Euro-North American) view of nature, to a large extent because of the influence of Darwin and those adopting and applying his ideas. Amongst these views, the idea that nature is diverse and that this diversity is necessary for evolution, that the natural processes are constant struggles and that 'nature' as well as species are always changing, is the most fundamental. In addition, it was also realized that there was a balance in nature that made it function in a regulated and predictable manner despite these Darwinian processes (Botkin 2012). It was also becoming clear that ecosystems were being lost and species threatened. This new paradigm, was rapidly applied to the Galapagos, most notably by the scientists of the California Academy of Science in 1906. Concern was now being expressed that the island's remarkable animals and plants, such as the giant tortoises, were being decimated by the local inhabitants and that hunting, habitat change and harvesting of animals and plants was threatening the very species that were so important for Darwin's theories (James 2017).

Connections between humans and nature were also explored in many different ways. In Germany, 'natural philosophers' such as Alexander von Humboldt and Ernst Haeckel (1834–1919; https://en.wikipedia.org/wiki/Ernst_Haeckel#Embryology_and_recapitulation_theory) had already argued that there was a connection between human wellbeing and natural areas. These views challenged the very foundations of western thought and the view that humans are different from nature and the anthropocentric view of humans as masters of the natural world. In the 19th century, economists such as John Stuart Mill (1806–1873; https://en.wikipedia.org/wiki/John_Stuart_Mill) and William Stanley Jevons (1835–1882; https://en.wikipedia.org/wiki/William_Stanley_Jevons), were also warning that economic development could not go on indefinitely because mineral and land resources were being depleted. Darwin's views on this subject were discussed in his second iconic work, *The Descent of Man*, published in 1871. He argued—although today we might say "demonstrated"—that humans are connected to nature not only in a biological and morphological way, but also in terms of behavior, including 'morals', by claiming that evidence of our evolutionary past is reflected at many different ways. As a result, this 'Darwinian revolution' meant that as humans could be demonstrated to have were evolved from *other* animals, and our behavior could also be understood as a development of animal instincts.

In the early, mid 20th century, scientists such as the Austrians Konrad Lorenz (1903–1989; https://en.wikipedia.org/wiki/Konrad_Lorenz) and Irenäus Eibl-Eibesfeldt (1928–2018; https://en.wikipedia.org/wiki/Iren%C3%A4us_Eibl-Eibesfeldt), and Dutch Nikolaas Tinbergen (1907–1988; https://en.wikipedia.org/wiki/Nikolaas_Tinbergen) took these ideas even further and created a concept of animal and human *ethology*. This view, which demonstrated a connection between humans and animals, challenged the dualistic view of humans versus nature that was a key concept of traditional western Judeo-Christian thinking. Another link between Darwinism and the social sciences came from the influence of liberal ideas about self-regulating systems that started with proponents such as the pioneering economist Adam Smith (1723–1790; https://en.wikipedia.org/wiki/Adam_Smith), famous for his seminal work *The Wealth of Nations* published in 1776 (Sandelin et al. 2014; Van de Haar 2015; Harnhart 2007) and extended to biological thought and then came back to social science as the study of self-regulating social systems.

This Darwinian tradition was part of the influence for the conceptual changes at the end of the 19th century that resulted in some of the earliest efforts to create conservation parks and protected areas to assure that the survival of some of the most basic facets of nature could be guaranteed. The development of the sciences of field biology, conservation biology and evolutionary studies also framed the establishment and development of the national parks globally, to guarantee the continuation of the natural processes of nature, including evolutionary, free from interference from humans. In must not be forgotten, however, that although the term 'National Park' now typically implies large designated areas of, relatively, unaltered natural space, the term was first formally applied in legislation in 1879 to an urban public open space in Sydney, in the colony of New South Wales, Australia (Griffiths and Robin 1997).

From their first establishment, such areas became prime places for not only for nature tourism (including safari), but also for scientific study and conservation. Scientists such as Darwin and Humboldt—and others—had established travel as an intellectual metaphor for exploration, but with more modern methods and facilities new generations of urban-residents could now travel with less peril and more comfort. Very soon, tourism became a major source of funding for the new parks …

In the United States, an eco-centric view of nature emerged based on the ideas of campaigners such as Henry David Thoreau (1817–1862; https://en.wikipedia.org/wiki/Henry_David_Thoreau) and John Muir (1838–1914; https://en.wikipedia.org/wiki/John_Muir) resulted in the establishment of the first national *conservation* parks in the world, with the USA leading the process of designating natural areas for preservation. In 1860s parts of Yosemite were already protected (Spence 1999), but the first formalized 'park'—Yellowstone National Park—was created with the approval by the US Congress in Wyoming in 1872 as a: "… *public park or pleasuring ground for the benefit and enjoyment of the people*" and set apart from any type of productive or urban use. As discussed further below, this legislation removed more than two million acres from use by its indigenous inhabitants … (Spence 1999).

These earliest protected areas primarily celebrated landscapes and the beauty of nature (Ise 1961; Runte 1979;

Everhart 1983; Beinart and Coates 1995; Pritchard 1999), still reflecting, in part, a romantic view of nature as immortalized in the paintings by famous artists. In France, for instance, the ancient forest of Fontainebleau near Paris, which had inspired many artists including the Barbizon School, was formally protected as a 'nature reserve' as early as 1861, but primarily for artistic reasons (but with a history of campaigns to protect its old growth trees going back to at least 1836); de Wever et al. 2015. Other examples include Yosemite National Park and the Grand Canyon in the US.

Initially key movers such as the artists George Catlin (1796–1872; https://en.wikipedia.org/wiki/George_Catlin), and John James Audubon who held a romantic view of landscape and native people argued that preserving nature must also include indigenous peoples (Spence 1999). Catlin has long been considered to be the patriarch of an intellectual genealogy that includes Henry David Thoreau, John Muir and Aldo Leopold. However, contrary to the later thinkers, Catlin's view included the presence of indigenous inhabitants (Spencer 1999). This view and ideology later changed dramatically with people such as Frederick Billing (1823–1890) (Spence 1999, p. 11) who was involved in the creation of Yosemite in the 1860s—and later as director and then president of the Northern Pacific Railroad, played an important role in the creation of Yellowstone. Billing and others such as Samuel Bowles, wanted to preserve areas such as these and Big Horn, not only for people to enjoy, but also so they would become symbols of a young and expanding nation. They believed that the different first nations in the area were just vanishing and could be ignored, and recommended to Congress that native people be removed from these wild areas (Spence 1999)—a clear case of 'ethnic cleansing' in the name of conservation (Kantor 2007). As a result, the March 1, 1872 Yellowstone Park Act removed more than two million acres from settlement, occupancy or sale, but in so doing Congress also ignored existing treaties with different nations (Spence 1999) (Fig. 4.2).

Hence, the creation of the concept of national parks began in the USA with the conservation of large areas of what was considered to be 'wilderness' or 'untamed nature' (the original meaning of wilderness in old English is simply a 'land inhabited only by wild animals'); https://en.oxforddictionaries.com/definition/wilderness. Under this problematic construct that denies human intervention in nature, even by indigenous hunters and gatherers who may have interacted with the same nature for 1000s of years (Cronon 1995), national parks such as Yellowstone and Yosemite where established. In the case of Yellowstone, the original inhabitants had shaped the environment through hunting, fishing and gathering, the use of fire, and other activities since the time of the 'paleoindians', the earliest cultures recognized in the region (dating to at least 12,000 years ago with the Clovis people). A view was nevertheless constructed that large areas of the American West were people-free and hence suitable for the creation of national parks within which to preserve an idyllic 'wilderness'. This view was, and indeed still is in many places, based on a 'western' prejudice that urban occupation must, almost by definition, included permanent constructed settlements and monuments. As discussed by Wilson (2005), however, the interaction of nomadic groups is far subtler and much less invasive, and hence easily overlooked—even ignored—under such a view.

Fig. 4.2 Yellowstone National Park, US **a** Bison are still roaming free in this protected area. *Photo* D. F. Kelley; **b** view of Yellowstone Falls looking up the Grand Canyon of the Yellowstone—this is one of the most popular tourist viewpoints in the park. *Photo* D. F. Kelley

Alston Chase (1987) has documented some of these tensions in the conservation of Yellowstone as managers tried to suppress fires, the activity of indigenous people and create what *they* conceived to be an idyllic landscape. Chase mentions that the presence of the Shoshone, Crow, Bannock, Blackfoot and Sheepeater nations was systematically disregarded and indeed it was believed that indigenous peoples were inevitably in the process of disappearing, linked to the expansion of European colonization, settlement and their systematic 'removal' from many areas (Spence 1999).

Pritchard (1999) discusses the history of nature resource management in Yellowstone National Park as a dispute between a perspective that considered that the intervention of the European colonizers was necessary to the proper management of wildlife within the National Parks versus another one that considered that the newcomers were not required to intervene in the established natural systems. This dispute remained important through the twentieth century, however, as the influence of scientific opinions on such debates increased. As science became a more important way of defining nature management, its implementation became increasingly professionalized. Themes which emerged as key issues included the regulation of ungulate and predator populations and the recognition of disease as a factor shaping the population of animals became key issues (Botkin 2012). There was also an increasing recognition of the damage that hunting both by settlers and the indigenous peoples was having on some of the wildlife.

The top down approach to the creation of protected areas such as National Parks was the result of the Eurocentric and ecocentric view of nature and a construction of 'wilderness' that needed to be managed by a government in a top down fashion. This concept of mas areas without a history of human intervention certainly parallels the development of medieval royal and aristocratic hunting 'parks', where local populations were excluded to 'conserve' the resource for a privileged few. These views of course ignored not only the established presence of humans in such areas but also the role that such communities would have had in shaping the 'valued' environment. In a somewhat contradictory way it was now considered that somehow these wild areas actually had to be *managed* and the paradoxical view thus created in the early part of the 20th century, was that in order to protect this cultural construction of untamed wilderness, more intervention was needed.

The early history of national parks everywhere in the world has also tended to demonstrate little concern for the interests of local or indigenous people. Most often, in the US, India, many parts of Africa and some in South America, communities were forcibly removed from the designated areas in order to make way for a new managing 'elite' and recreational tourism, or tolerated only as a paid labor force (Caruthers 1989). Fortunately, the management of national parks as 'fortresses' that exclude local people, has now evolved in many areas into a broader 'community conservation' in which all sectors of the population have a stake—perhaps best expressed within the concept of Biosphere Reserves and Geoparks, a discussed below. However, the expulsion of local people from designated areas still continues in some countries, despite representing a now widely discredited anachronism, including reported cases in Kenya, Uganda, Bangladesh, Namibia, Botswana, Cameroon, Ethiopia, South Africa, Argentina, India, Chile, Sri-Lanka, Thailand, Tanzania, Brazil and even Ecuador (*The Guardian*, 2016, see https://www.theguardian.com/global-development/2016/aug/28/exiles-human-cost-of-conservation-indigenous-peoples-eco-tourism).

Similarly, in Europe, the vision of 'parks' as areas that have to be preserved for the sake of nature or for the sake of science was strong and exclusion of local communities from using any an area's resources, including simply as traditional grazing land, has also locally been a phenomenon. In Europe, scientists and scientific organization were often been a major source of pressure to establish national parks, for instance in Sweden and Switzerland which created the first national parks in the region in 1910, national academies of sciences were deeply involved in the process. Whilst in Sweden scientists became more and more sidelined by touristic interests, in Switzerland they have maintained a dominant role in the conceptualization and realization of the national park and other Alpine countries followed (e.g. France, Italy, Germany, Austria, and Slovenia) and initiated national park projects modeled after the Swiss example.

Elsewhere in mountainous areas, such the Spanish Pyrenees, other national parks have been established. The National Park of Odessa and Monte Perdido (https://www.mapama.gob.es/es/red-parques-nacionales/nuestros-parques/ordesa/), designated in Aragón, Spain, in 1919 is a good example and was primarily established to help protect the last remaining population of the *buscardo*, *Capra pyrenaica pyrenaica* (Schinz) an endemic Pyrenean subspecies of mountain goat. Although the extinction of the subspecies in 1992 has been considered to be a 'mystery'—competition with other species, illegal hunting and exotic diseases have been blamed (https://en.wikipedia.org/wiki/Pyrenean_ibex)—there is perhaps a strong link between the removal of grazing by livestock by local communities. This traditional 'human intervention', had controlled the expansion of forest, and maintained open pasture at lower levels which the buscardo would also have needed to survive long mountain winters (Fig. 4.3).

Support for nature conservation and the creation of national parks in many European countries decreased inevitably after 1914 with the onset of the First World War and its aftermath. In Switzerland and other countries, designated trails were established to control the movement of people and some protected areas were entirely closed to the

Fig. 4.3 A typical barrage of sign boards greeting a visitor to a National Park, emphasizing what they cannot do, rather than what they can; Parque Nacional de Odessa y Monte Perdido, Pyrenees, Aragón, Spain. *Photo* K. N. Page

public, which was also the rule for French nature reserves. Such political moves—often 'scientifically' justified—however, violate the basic principle under which national parks were established as places for public enjoyment of nature. The process of creating areas of conservation, sometimes without adequate scientific justification, and the associated laws and regulations perceived necessary to manage—or in many cases 'control' them—inevitably led to the criminalization of many of the activities traditional practiced by local communities. For instance, subsistence hunting now became 'poaching' and gathering fire wood, criminal damage or theft. In many areas of the world, this style of conservation agenda pushed by scientists and pressure groups from 'developed' countries, reflecting their own cultural perspectives and prejudices, and often scant regard for the original inhabitants, indeed historical 'owners,' of the land to be designated and controlled.

Somewhat ironically, given its colonial heritage, it was perhaps the UK that first developed a more inclusive concept of National Parks in 1949, as part of a socialist vision for the re-building of the country after the devastation of the Second World War. The 1949 National Parks and Access to the Countryside Act created the legal framework for the designation of England, Scotland and Wales first nature conservation areas, including National Nature Reserves, 'Sites of Special Scientific Interest' ('SSSIs'), and crucially, National Parks—and a state body to implement and oversee these designations, in the first instance the Nature Conservancy (Evans 1997). Fundamental to these new 'National Parks' was public access, as most unfenced countryside, such as mountain and moorland, in England and Wales was (and much still is) owned by a small number of hereditary aristocratic estates. Most of these estates actively excluded people from their land, which had led to mass-protests in the 1930s, for instance in the uplands of northern England (https://en.wikipedia.org/wiki/Mass_trespass_of_Kinder_Scout), as a modernizing society's tolerance of the old order began to wane.

Although, IUCN's classifications of protected areas (https://www.iucn.org/theme/protected-areas/about/protected-area-categories) somewhat patronizingly assign these areas to 'Category V' as no more than protected landscape areas and not 'true' National Parks—something which has been strongly challenged in the UK—they certainly have a much closer cultural connection to a nation than areas in other countries where governmental and other authorities strive to restrict public use, even prevent it. Within UK National Parks public access to open countryside is now effectively a legal right, but crucially no communities are displaced and traditional activities such as farming and other economic activities, such as forestry, as were in place at the time of designation, continue. The legislation, however, does aim to prevent any changes that would damage or destroy natural landscapes, associated habitats and features of cultural importance—and management authorities in-place also strive to restore damaged areas to a more 'natural' and/or 'traditional' state. In addition, all National Parks in the UK have a level of democratic control and elected councils, again helping ensure that the community is part of the decision making process. The result is something that could genuinely be called a *National* Park … (Fig. 4.4).

Once it was consolidated in the USA, however, the concept of creating national conservation parks to the Yosemite-Yellowstone model was exported to other areas of the 'developing' world such as Africa, Asia and Latin America in the first half of the 20th century (Harroy 1972, p. 9; Nelson et al. 1978). Africa was one of the earliest places outside of North America, Europe and Australia where such parks became established. Here, as in North America, the arrival of European settlers created a frontier economy that lead to the local elimination of much of the native fauna and flora, especially in the 19th century. New technologies such as firearms, medicine, railroads, and markets for wildlife products, as well as the conversion of land to agriculture, only increased the trend towards the elimination of native wildlife. A combination of habitat loss and overhunting created both local and total extinctions of some species, such as the bloubok and quagga in the Kruger area of South Africa (www.kruger2canyons.com). This wholesale decimation of populations of large mammals and concerns about their possible extinction had already led the Cape Colony to protect elephants and buffalo in 1858, and the 1886 Cape Act for the Preservation of Game was the first systematic conservation legislation in Africa. This framework laid the foundation for a centralized control network

Fig. 4.4 Dartmoor National Park in Devon, SW England. Traditional grazing activity on an upland periglacial landscape. *Photo* K. N. Page

for protected areas and the conservation of wildlife (van Heijnsbergen 1997) and the first national parks in the region were established in the late 19th and early 20th century in the Cape Town area and Nathal region, with the national South African National Parks body being established in 1926 (https://en.wikipedia.org/wiki/South_African_National_Parks).

The proposed 1900 London Convention Concerning the Preservation of Wild Animals, Birds and Fish in Africa attempted to establish some level of international agreement between European and African nations to promote such conservation on a much broader scale. Although initially having little effect, the original document was revised and eventually signed as an international convention in London in 1933, known as the *Convention Relative to the Preservation of Fauna and Flora in their Natural State* (or just the 'London Convention'), and this time it was successful in creating a system to protect the wild life across parts of the continent (IUCN 2004; https://en.wikipedia.org/wiki/Convention_Relative_to_the_Preservation_of_Fauna_and_Flora_in_their_Natural_State). These conventions were instrumental in the establishment through the first half of the 20th century of a number of large nature reserves across southern Africa, including Kruger (South Africa), Hwange (Zimbabwe), Chobe (Botswana), South Luangwa (Zambia), Etosha (Namibia), Serengeti (Tanzania) and Tsavo (Kenya) (Neumman 1998; Howkins et al. 2016).

There remained, however, a clear association in Asia and Africa between the expansion of national conservation parks and related initiatives with colonial rule and an elite perspective on nature. One of Africa's first national parks, for instance, the Albert National Park in the Belgian Congo was established in 1925 with the support of Belgium's King Albert. The overriding conservation objective of the park was the protection of the mountain gorilla (*Gorilla beringei beringei* Matschie), the rare and charismatic primate highly threatened by extinction, and scientific research was declared the park's main goal and tourism was discouraged. Such views were also typical across the National Parks of the former British Empire where the primary purpose of such areas in Africa was not entertainment but scientific conservation. Nevertheless, science was not always seen as important for the *management* of conservation parks in the beginning and in many areas such as Africa, it did not play a significant role in the process until the second part of the twenty century (Carruthers 2007). Indeed, despite stated nature conservation aims, the initial product of some such designations, was still effectively a game preserve for recreational hunting by colonial elites (Marks 1984; MacKenzie 1988).

Inevitably, however, as in North America, indigenous and resident communities, such as the Massai in Tanzania, were removed from their ancestral territory to make room for these reserves, for instance in the early 20th century to create national parks such as the Kruger. Crucially, the creation of this park came at a critical moment in the creation of the nation, and hence it served as a unifying symbol for white identity (Caruthers 1989; Neumann 1998). A similar situation occurred with the Serengeti in British ruled Africa (Neumann 1998). In areas such as Kruger, fences were placed to keep the original inhabitants out of the new protected area, although not all such initiatives were governmental led... For instance, the London based *Society for the Preservation of the Fauna of the Empire,* which was instrumental in establishing a system of parks in Tanzania from the 1930s to the 1950s, was able to take over territories in Africa for both game hunters and nature preservation (Neumann 1998). In some cases, however, these initiatives remained strongly scientifically influenced, as state and independent scientific organizations, such as Paris's renowned *Muséum national d'Histoire Naturelle,* took charge of areas and established conservation measures. In 1927, for instance the MNHN established ten *Reserves naturelles intégrales* in Madagascar dedicated to the sole purpose of scientific research—but facilitated by the colonial power France then had over the territory and inhabitants of the island (Neumman 1998; Howkins et al. 2016).

By the later 20th century, however, many of these designated national parks had become important attractors of tourists, and such activities now fund a remarkable 75% of their management activities (https://www.sanparks.org/about/history.php). An interesting relationship has now developed in South Africa, for instance, where alongside state national parks with well-developed touristic infrastructures, private game reserves flourish where a modern economic, rather than aristocratic, elite can continue to recreationally kill large mammals, without, in theory, decimating natural populations. Part of the income for the state parks, however, comes from selling 'surplus' animals to the hunting parks, for instance when issues of overcrowding develop, hence, effectively supporting the continuation of tourism based on trophy hunting (Lindsey and Romanach 2007) (Fig. 4.5).

The net result, however, is much of the country's iconic 'wildlife' now only thrives within the fences of the state parks and game reserves and elsewhere the continued expansion of human societies and farming has virtually eliminated many species along with their habitat. Nevertheless, countries such as South Africa has always had a different socio-cultural vision for conservation than Europe or America (Anderson and Grove 1987), often driven by a rural colonial elite, rather than an urban, economic and political constituency.

In Latin America, the creation of protected areas, and hence National Parks, however, has been mostly a process of modernization and assertion of national territory on the part

Fig. 4.5 A view of Addo Elephant National Park in South Africa—established when the extermination of African Elephants by orange growers led to concerns that the species would be eliminated from the region. From an initial 11 surviving elephants in 1931, the population now numbers around 600 (https://www.sanparks.org/parks/addo/tourism/history.php). The entire area is surrounded by a very strong fence. *Photo* K. N. Page

of more recently established nations. Currently in Latin (i.e. south and central) America there are some 120 national parks in 17 countries covering 18.5 million ha. One of the first to be established across the region was the El Chico forest reserve in Mexico in 1898, and although subsequently renamed as the Parque Nacional El Chico, it is still considered to be the nation's first national park (http://www.parqueelchico.gob.mx/).

Argentina established its first national park with a donation of 7,500 ha of land by the explorer and academic, Francisco P. Moreno (1852–1919; https://en.wikipedia.org/wiki/Francisco_Moreno) in 1903 which became the basis for the "National Park of the South" in 1922—and part of the Nahuel Huapi National Park from 1934, now covering some 785.000 ha. The reasons for designating this area are largely taken from the original USA concept, with scenic beauty, and iconic animals being important criteria its creation. Argentina also created a Direction of National Parks and in 1934 established the Iguazu National Park and three years later the Lanin, Puelo, Los Alerces, Perito Moreno and Los Glaciares national parks (https://www.parquesnacionales.gob.ar/). Chile established the adjoining 'Vicente Perez Rosales National Park' in 1926, in the scenic 'South Andean Lake District' (http://www.conaf.cl/parques-nacionales/parques-de-chile/). Ecuador followed with the Galapagos Islands National Park in 1934, and Brazil and Venezuela established their first parks in 1937.

As in other parts of the world, in Latin America the issues around the inclusion or not of indigenous people has divided perspectives about the natural parks. In Brazil some of the early promoters of the concept of national parks such as Cândido Rondon (1865–1958; https://en.wikipedia.org/wiki/C%C3%A2ndido_Rondon) and the activist Villas-Bôas brothers, (Orlando (1914–2002), Cláudio (1916–1998) and Leonardo (1918–1961); https://en.wikipedia.org/wiki/Villas-B%C3%B4as_brothers) believed that the indigenous inhabitants were the most appropriate custodians of the forest. For the Villas-Bôas brothers, the protection of these indigenous peoples was the best insurance for nature conservation of nature against non-indigenous intruders (Howkins et al. 2016). Rondon also wanted to see a combination of indigenous protected areas and national parks. As a result, the creation of several national parks in Brazil in the 1950s and 60s were ground breaking in this changed perspective. However, even where indigenous people were allowed to remain they often had little say in the management of the area, becoming little more than part of a cultural landscape. One concept that was imported to Brazil at the beginning of the development of the national parks was *Monumentalism*. Examples start with the first four Brazilian parks: Itatiaia (1937) was created around a mountain peak believed at the time to be the highest in Brazil; Serra dos Órgãos (1939) contains a set of prominent peaks visible from downtown Rio de Janeiro; Iguaçu (1939) showcases the impressive waterfalls of the Iguaçu River; and the now defunct Paulo Afonso Park (1948) was also created around a major waterfall on the São Francisco River (Howkins et al. 2016).

Elsewhere across Latin America, the US model of national parks has been implemented and indigenous populations have been displaced, even latterly in Brazil, and hence the concept of national parks across the region has been a mixture of external and internal ideas. Nevertheless, the model of 'National Park' which typically persists in Latin America and globally is still based on the model first established in the USA, and then exported to much of Europe, Africa and beyond. Intertwined with concepts of 'untouched wilderness' and 'monumentalism', and a top-down management implemented by imported professional managers with an exclusion of local human activity—whether traditional or not—such areas continue to represent a paradigm which is becoming increasingly anachronistic and even scientifically, difficult to justify. Many examples are now available to demonstrate that collaboration and partnerships with local communities and even visitors, represents a model which is more compatible with modern democratic societies—it may even, ultimately, be more sustainable.

4.1.3 From Policing to Management

Initially the protection of many national parks, such as Yellowstone, management was more about policing, with park rangers, given legal powers to *enforce* the protection of nature, in a way dictated by the area's management authorities. Subsequently, however, increasing scientific awareness and the appearance of professional managers led to an evolution of this role from purely enforcement to informed management. When national parks were first established in the late 19th and first half of the 20th century, however, ecology was in its infancy and the only relevant field sciences were those relating to agricultural or veterinary matters. In areas of developing countries such as in Africa, for instance, environmental management often continued to rely on the culling of predators as a way of addressing any decrease in ungulate population (Howkins et al. 2016). As zoology and botany were still primarily studied at universities, with systematic field studies still uncommon, the task of early wardens and game rangers remained primarily policing, not scientific or habitat management work. By the late 1940s, however, biology had begun to widen its scope into environmental studies with the realization that it is important to study wild animals and plants in their natural ecological settings. As in the rest of the world, during this time in Africa there was the institutionalization of conservation efforts and the 'professionalization' of managers, as University-trained conservationists and biologists were hired. Such changes were promoted by a establishment of a series of European and North American based institutions and organizations such as the World Wildlife Fund (WWF) in 1961 (https://www.worldwildlife.org/about/history), which emerged to assist governments in the planning and management of these areas—hence supporting the development of scientifically informed conservation measures and related employment. As a result, simple predator control was replaced by more informed 'management by intervention' and the active manipulation of habitats and wildlife populations, including through land management practices—and more recently by the concept of systems ecology, resilience theory and local knowledge (Folke 2004).

After World War II, with the creation of global institutions such as UNESCO (The United Nations Educational, Scientific and Cultural Organization) in 1946 (https://en.unesco.org/; https://en.wikipedia.org/wiki/UNESCO) and IUCN (the International Union for the Conservation of Nature) in 1948 (https://en.wikipedia.org/wiki/International_Union_for_Conservation_of_Nature; https://www.iucn.org) it became possible to both promote the best informed conservation management practice to governments globally, as well as gain global support and recognition for national activities—not least through listing as World Heritage following the 1972 UNESCO convention. Suddenly, national nature conservation and protected area management—including the concept of 'wilderness'—could be objectively placed in a global context and guidelines established and disseminated for its informed management.

4.2 A History of Galapagos National Park (GNP)

4.2.1 Darwin and Collecting Nature in the 18th and 19th Centuries

In the Galapagos, as in other regions of Latin America, the expansion of the paradigm of conservation and of top-down, command and control strategies for conservation, the use of science and research and the professionalization of managers is a consequence of the belief that nature can be successfully managed if appropriate measures are taken. In the Galapagos, restrictions on activities that early conservationists felt altered the pristine environment were first imposed in the 1930s, including hunting, cutting of certain trees and fishing. And in from the 1970s, further restrictions were implemented in marine areas, culminating in 1998 with the creation of the Special Law of the Galapagos (LOREG—Ley Orgánica de Régimen Especial de la Provincia de Galápagos) and the Galapagos Marine Reserve (GMR). It is important, however, to understand the background to this legislation and the issues which led to what some might consider to be scientifically justified and necessary controls, but which others might consider draconian …

When Charles Darwin arrived in the islands, they were already under serious ecological threat with the archipelago having become a provisioning point for ships from many nations. Of particular focus of this activity were the giant tortoise populations, which could be easily 'harvested', and due to their slow metabolism, individuals could survive long periods in the hold of ships before being consumed. Reports indicate that some ships loaded as many as 400 tortoises into their holds for a single voyage, and it is estimated that up to 200,000 tortoises a year may have been taken and killed as part of this activity—leading of course to the well documented extinction of species on some islands (Pritchard 1996). Darwin's writings in the Voyage of the Beagle (1839) suggest that did he not seriously consider the inevitable effect that this removal might have on the tortoise populations of the islands—but he does note that tortoise meat was rather poor to eat, but young individuals made a tolerable soup. Indeed, Darwin's was already aware that the then Governor of the Islands had noted that the tortoises were already not so plentiful and had calculated that at the then rate of exploitation, 'supplies' would have run out within 20 years (Desmond and Moore 1991, p. 176), e.g. by 1864 … (Fig. 4.6).

But as with previous expeditions to the Islands, Darwin came with the desire to collect as many specimens of animals and plants as possible, as the study of nature at this time was still very much in the realms of documentation, with barely no consideration of conservation. One of the British botanists that classified most of the collection brought back by Darwin, Hooker, realised that although the flora had many close links with the plants of South America, his conclusions on the distinctiveness of the archipelago and individual islands were astonishing. He was surprised that of the 217 species collected, 109 were endemic of the Galapagos and 85 of were found in a single island (Journal of researches 2nd ed., pp. 395–397.) Although the different shapes of tortoise carapace on different island would become important for Darwin's later theories, he only collected 30 tortoises on James Island (Santiago) and 30 on Chatam (San Cristobal). In addition, as most had been collected by the crew of the Beagle as source of fresh meet during the journey, most of the carapaces were thrown into the sea after the flesh had been consumed (Sulloway 1984). It is interesting to note, however, that Darwin, like many young men of his social group had been keen on hunting, although his passion diminished during his voyages (Steinheimer 2014; Sulloway 1984).

These pioneering collections and the contribution the fauna and flora of the Galapagos subsequently made to the development of Darwin's theory of natural selection and his iconic book, *The Origin of Species* (1859), strongly influenced later Western views of the Galapagos. Indeed, thanks to Darwin the species of the Galapagos, have become more than just endemic species, as they are a core part of the evidence Darwin presented as he completely changed views of the origins of nature and its relationship to humans. Thus, even in the later 19th century and the 20th Century, conserving the animals and plants of the Galapagos was conceived as much a tribute to Darwin and his incredible influence on the Western cosmology, as it was about conserving a unique ecosystem.

During the 19th and earlier 20th century it was believed that since the animals and the plants in the Galapagos were threatened by the activities of residents and visitors, it was necessary to collect them to preserve them for future generations. With few exceptions, many of these early expeditions were not guided by the larger of questions of biology and ecology, nor was fieldwork guided by theory, the aim was primarily to collect. Notable examples from the late 19th Century and early 20th are the intensive collecting expeditions sponsored by the British millionaire Sir Walter Rothschild and the California Academy of Science (CAS). The CAS scientists, in particular, perceived the possibility of the imminent disappearance of the native tortoises as local people were killing them to extract oil and hence they collected 266 tortoises and many specimens of birds, reptiles, insects, mammals and plants. Yet any desire to *conserve* the animals that were disappearing cannot explain why, ultimately, they collected more than 78,000 specimens (James 2017).

Fig. 4.6 Giant tortoises on the Galapagos—a 19th century vision (note the sailor hunting with spear and rope …)

4.2.2 From Collecting to Conservation

The desire to collect was justified by the view that as many of these animals and plants were unique to science, collecting was essential for their study. As with many scientists at the end of the 19th and beginning of the 20th century, this approach was linked to a realization that species could become extinct within a human timescale and not just geological. The vertebrate palaeontologist, Georg Baur (1859–1898; https://en.wikipedia.org/wiki/Georg_Baur), however, is a rare example of an early scientist who returned from the archipelago in 1891 concerned about the survival of species such as the tortoises that were being exploited by visitors and residents for their oil and meat (James 2017; Hennessy 2017, p. 70). The idea that it is necessary to collect specimens for academic and scientific centers, to literally 'preserve' them, was, however, gradually changing as people started to realize that it could be possible to protect them in situ, i.e. to *conserve* those species.

In the 1930s, field biologists started to call for action as they saw a increasing degradation of Galapagos environments (Hennessy 2017, p. 72). The idea that the Archipelago was a natural laboratory for the study of evolution was also consolidated during that time, and between 1930 and 1950, North American and European naturalists and scientists campaigned to create a research station in the Galapagos. They lobbied politicians and other key figures in Ecuador and elsewhere and although they gained support from parts of the Ecuadorian government and international institutions such as UNESCO, IUCN and WWF, initial campaigns failed to establish the legislation necessary to create a Science Station (Hennessy 2017). In 1935, the American explorer and anthropologist, Victor Wolfgang von Hagen (1908–1985; https://en.wikipedia.org/wiki/Victor_Wolfgang_von_Hagen) erected a monument on San Cristobal as part of his effort to protect the archipelago in Darwin's name; his plan was to have the approval of the Ecuadorian government, but for non-Ecuadorians to manage the station

(Hennessy 2017). However, as von Hagen was not considered to be a scientist by the scientific establishment of the time, he did not gain sufficient support and his efforts failed to convince Ecuadorian bureaucrats to establish a national park—and WW2 brought his efforts to a standstill. After visiting the islands on a marine expedition in 1950s, Austrian Scientist, Irenauss Eibl-Eibesfeldt (1928–2018; https://en.wikipedia.org/wiki/Iren%C3%A4us_Eibl-Eibesfeldt) and Canadian-American biologist Robert Bowman (1926–2006; http://biology.sfsu.edu/people/robert-bowman) also wrote to IUCN expressing concern about the threats to the Galapagos' iconic species (Hennessy 2017).

Scientists such as Bowman were concerned that the Ecuadorian authorities did not understand that people visited the Galapagos because of the environment and the unique animals that can be found there, but that these are the same animals that the inhabitants were hunting and fishing, potentially to extinction. Notably, Ecuador had also funded expeditions to the Galapagos, such as that by the Darwinian Scientist, Theodore Wolf, who had stressed the importance of the islands for understanding natural sciences (Sevilla 2016). As a result, in the 1930s the idea of conserving the Natural Laboratory began to gain acceptance both among the international community of scientists and among Ecuadorian authorities. Robert Bowman, in particular, as well as other scientist and Ecuadorian officials were beginning to argue that the promotion of tourism in the Galapagos would be an effective way of educating the local people and, hence, help protect the unique wildlife (Hennessy 2017).

The idea that islands could become a tourist destination was not new, however, and in the 1920s and 1930s accounts of the Galapagos such as those written by American naturalist, William Beebe (1877–1962; https://en.wikipedia.org/wiki/William_Beebe) and other writers encouraged affluent Europeans and North Americans to not only to visit the islands by luxury yacht, but also to stay in the Islands. A significant milestone in this process was the visit of the two respected biologists, Eibl-Eibesfeldt and Robert Bowman in 1959, accompanied by a photographer and an illustrator for LIFE magazine. The article that they produced which was published on the September 8th 1959, as "*The Fantastic Galapagos: Darwin's Treasure of Wildlife*", captured the imagination of readers in the North America and Europe and set the stage for the future conservation of the Galapagos and for the large number of tourists that eventually were to fuel the economy of the islands (Hennessy 2017). It was at this time that these scientists started to look for a place to establish the proposed research station on Santa Cruz and the Congress of Zoology in London in 1958, adopted a resolution supporting the creation of an international committee that would regulate its conservation activities within Ecuadorian territory (Hennessy 2017).

This new attitude was the result of the realization around the world of the importance of protected areas where endemic species could be conserved, and the institutionalization of these ideas after WW2 created new organizations which could promote this goal, such as IUCN, UNESCO and WWF. Motivated to a large extend by scientific interest in the Galapagos, it became established that the region needed to be protected from any threat to its conservation. There was a sense that the conservation of the archipelago would consist of maintaining a pristine, untouched and perfectly balanced system, as free as possible from any human influence. It was at this time that the strong relationship between science and conservation was established, and following many of the development discussed above, it was considered that the local population was an significant threat to the survival of the endemic and unique species, and needed to be controlled.

As a result, in 1959 the Galapagos National Park (GNP) was created as the result of a combination of the efforts of people from developed nations and some Ecuadorian authorities. Scientists, most of them strong supporters of Darwin's ideas, considered that it was essential for the Galapagos to be promoted as a place to celebrate Darwin and study his ideas, at a time when Darwinism was being consolidated internationally as the dominant paradigm in the life sciences. The Galapagos was seen not only as a natural laboratory but also as a shrine to Darwin and his ideas—hence it is no coincidence that key supporters of the establishment of the Galapagos as a protected area, were important contributors to the neo-Darwinian synthesis such as Huxley, Bowman, Eibl-Eibesfeldt and the British evolutionary biologist David Lack (1910–1973; https://en.wikipedia.org/wiki/David_Lack). These conservation scientists were the product of their own time, merging views of nature coming from a Darwinian scientific tradition with romanticism fused into a conservationist strategy that sought to protect and maintain nature as unique and pristine. In addition, for the Ecuadorian authorities, the creation of the GNP, as a result of European and American influence, was also seen as an important part of the narrative of Ecuador becomming a modern nation in tune with the scientific developments of the time.

The Galapagos-Darwin myth that was consolidated in the mid-20th century consists of two key aspects. Firstly, according to the myth (Sulloway 1984), Darwin discovered evolution on the Galapagos through direct observation, as it implies, or even states, that there was an immediate transformation of Darwin's thoughts based on the observation of the different species of Galapagos finches. The second aspect of the myth is that Galapagos was a pristine environment at the time of the arrival of Darwin. Thus, Darwin's visit became a base line for conservation across the islands, with the creation of the Charles Darwin Foundation (CDF) in 1959 being another step in the institutionalization of this narrative (Hennessy 2017). The Foundation and the GNP were created to celebrate 100 years of the publication of the Origin of the Species in 1859 and 150 from the birth of Darwin in 1809—and the creation, on paper at first, of these two organizations was a key step in the establishment of a strategy to conserve the key species that were living prove of the Darwin's theories. However, as had been appreciated from the 1970s and 80s, Darwin did not come to a eureka-type transformation while in the Galapagos,—and that he missed the importance of the finches until his return to England (Sulloway 1984) as it was the plants which had most interested him initially (Nicholson 1987). Nevertheless, the strong association between the discovery of the principles of biological evolution and Darwin's visit to the Galapagos has become part of the narrative of western science (Quiroga 2009; Hennessy 2017). Nevertheless, the creation of this myth was very important to secure the initial funding needed to support the conservation efforts in the Galapagos.

4.2.3 Conservation, Communities and the Establishment of the Galapagos National Park

Although early efforts to manage the Galapagos had little if any input from the local population, the number of residents in the Galapagos had been growing steadily as the area became better known as a destination for scientists, and later tourists. The islands had also become well known in mainland Ecuador as a destination for colonists and even former prisoners who wanted to start a new life. Although there had been visitors ever since their first discovery in 1535 by a Spanish ship, blown off course whilst navigating the coast of South America, the islands were not systematically colonized until the 1830s by the newly founded state of Ecuador. Within 30 years, plantations had been established on San Cristobal and then on Isabela, and when Darwin arrived in the islands in 1835, he describes between 200 and 300 local residents on Floreana. Even as late as the 1970s, less than 2000 people lived across the islands. The Ecuadorian government, however, established incentives in both Amazonia and in the Galapagos to encourage colonization, viewing these areas as frontiers where people could move and establish cattle and cultivate crops. Meanwhile, and contrasting this vision of the Galapagos, a number tourism studies were being carried out by the Charles Darwin Foundation. Tourism was seen by the research station as a less 'damaging' option than farming, and by using large cruise ships, tourists could be taken to different islands without any significant involvement from the local populations. This vision, therefore, would mean that the local towns and businesses would not be developed in economic and demographic terms, in sharp contrast to a modern view of sustainable development linked to social and community values.

The Charles Darwin Foundation was established as part of an effort to celebrate and continue the work of Charles Darwin and to protect the unique flora and fauna of the Galapagos. As discussed previously, a major theme was the belief that Darwin´s paradise could only be protected by foreign scientist and institutions, as the local Ecuadorian people who were destroying it. Conservation was conceived as the recovery of the island's ecosystems as closely as possible to an 'original state'—whatever this might mean. So in July 1959, the Ecuadorian government finally issued an executive decree establishing an emergency law that declared the Galapagos to be a protected area and a National Park—and five years later in 1964, the Charles Darwin Research Station was officially dedicated. The ceremony that took place brought together many of the Americans, Europeans and Ecuadorians who had worked to make the National Park and the Center a reality, including representatives of Ecuador's then ruling military junta then governing Ecuador The latter awarded Von Straelen, Bowman and other scientists, medals of honor. Crucially, an agreement was signed between the Ecuadorian government and the Charles Darwin Foundation confirming that the latter would serve as scientific advisor for the Galapagos National Park for the next 25 years. The new National Park was presented as a potential nature tourism destination and a source of revenue for the state through the creation of a profitable tourism industry. Scientists within the Charles Darwin Research Station, however, were expected to dedicate their time to administrative issues and a research agenda related to its advisory role (Reck 2017; Hennessy 2017).

To a large extend the creation of Natural Parks is the history of the creation of defined zones that are considered 'natural' and 'pristine' and establishing barriers or frontiers between these and other regions that can be seen as 'appropriate' for human settlement, where people can engage in agriculture and other aspects of land use and management (Valdivia et al. 2014). In the Galapagos, terrestrial zones and areas were first defined in the 1960s and 1970s, when the land was partitioned and areas were set aside solely for conservation. The creation of these borders inevitably generated tensions and problems between the resident local population and the managers of the Galapagos National Park.

Borders territorialize different visions, and as Valdivia et al. (2014) have stated, 'borders' in the Galapagos, as elsewhere, are more than lines on a map, they are continuously changing through lively, continuous encounters and the circulation of narratives and are constantly being negotiated. The increased presence of introduced species has meant that the limits and borders that differentiate terrestrial park areas from agricultural areas are not real ecological borders. Similarly, with the creation of the Galapagos Marine Reserve in 1998 and the zoning of coastal areas, space and activities were established and limited to specific areas. This zoning resulted in constant conflict as oceanic boundaries are not always enforced and fishing often occurs in areas that have been delimited for tourism or new areas are assigned for tourism that fishermen considered belonged to them. Indeed, both fisherman and tour operators 'blur' these boundaries which often become quite 'permeable'. As borders define access and produce power relationships, in the Galapagos, these borders represent different world views and divide not only space but visions.

The establishment of these boundaries and limits means that behaviors also have to be controlled and disciplines established. The regulation of such behaviors in the Galapagos National Park, led to the criminalization of activities which in the past used to be considered normal and even desirable on the part of the local people (many of whom were descended from the first colonists and considered themselves to be the real owners of the islands). Yet now they found themselves thrown out from their traditional territories both in land and in the sea.

Previously the islands residents would consume giant tortoise and marine iguanas, and kill the hawks that they considered a be a threat to their chickens and other domesticated animals. Sharks and marine turtles were sold to the Japanese boats that used dock in San Cristobal. Now they found themselves portrayed as the 'saboteurs' of the great works of the GNP and the CDF (e.g. in Nicholson 1987, pp. 124–5). Over time, protected and productive areas have become more differentiated as more and more areas are delineated as protected. At the same time, new regulations have been imposed, creating tougher sanctions for those who break the new rules—one of the most affected activities being one of the mainstays of the islands economy, fishing.

4.2.4 Fishing, Communities and Conservation

During the middle of the 20th century, the fishing industry in the Galapagos had started to grow and had become increasingly connected to a global commercial system. Until the 1950s and 1960s, large tuna boats fished different species including yellow fin (*Thunnus albacares*), skipjack (*Katsuwonus pelamis*), and albacore (*Thunnus alalunga*), initially mostly under foreign flags, but later mostly Ecuadorian. Fishing close to the shores of the islands ultimately decimated the stocks, whilst producing large amounts of ecologically destructive 'by-catch'. Within a few years, different local fisheries emerged, initially for Galapagos grouper (*Mycteroperca olfax*) and other types of demersal fish such as Misty Grouper (*Hyporthodus mystacinus*) that was exported to the mainland, and later spiny lobster (*Panulirus penicillatus* and *P. gracilis*) that was exported both to the mainland but also to other countries. In the early 1990s, with the increase demand for sea-cucumbers (*Stichopus fuscus*) from South East Asian countries, these echinoderms were intensively gathered and exported. As conservationists and the Galapagos National Park tried to limit the capture of these marine organisms the regulations applied and the policing implemented became the source of many conflicts with local populations which climaxed in the 1990s and 2000s (Quiroga 2013).

Inevitably, as a result of this intense fishing, driven by international demand, ideas about the creation of a Marine Reserve, as addition to terrestrial protection, started to circulate in the 1970s among conservationists and members of the Charles Darwin Foundation (Reck 2017). In the 1980s, some regulations were established to limit some fishing activities and try to regulate the activities within a marine protected area. Interestingly, the need to regulate external industrial fisheries, most of them based on the Ecuadorian mainland, unified conservationists, local fishermen and the tourism industry and in a united effort they pressed the Congress of Ecuador to establish the Galapagos Marine Reserve (GMR) in 1998. Despite the opposition from groups of industrial tuna fishers, the establishment of the reserve was finally approved by the Ecuadorian Congress in 1998.

Zoning of the new Reserve has been the result of constant negotiations between the different sectors: the government represented by the Galapagos National Park, science, conservation, tourism and the local fishing industry (Reck 2017). In this process more than 70% of the marine areas close to shore were assigned for local fishing, 7% to tourism and a similar area to science—the remainder, mostly areas around the ports, were left for multiple uses. These fishing areas were considered vital to sustain the livelihoods of the almost 1,000 local artisanal fishermen that were registered at the time (Fig. 4.7).

Fishing and agriculture as economic activities for the local population are now losing importance, however, as more people have moved away to the tourism sector-agricultural areas, in particular being abandoned. In some sense, we are now seeing a new transformation, related to the end of extractive and productive activities, as the economy shifts toward a service economy based on tourism. Many industrial fishing companies, however, who had originally opposed to the creation of the Galapagos Marine Reserve have now learnt to benefit from what is called the 'spillover effect'. As well illustrated by the Galapagos National Park map reproduced as Fig. 4.10, industrial fishing boats now exploit the ecological success and recovered

Fig. 4.7 Fish market in Puerto Ayora. *Photo* D. F. Kelley

fish stocks of the reserve area by maintaining a presence immediately outside of its boundaries to catch any straying outside.

Not surprisingly, local people have become increasingly frustrated that they were asked to sacrifice their own industry for the benefit of the large-scale commercial fishing and tourism industries that have benefited from the natural ecosystems of the Galapagos. They consider that the many outsiders involved in both trades are present in the Galapagos to make a profit without contributing, in any significant way, to the health and well-being of the resident population. Indeed, sometimes as a tourist in the islands, this sense of exploitation can be sensed, with an expectation of cash 'tipping' of the crew on some tour boats to levels even in excess of equivalent western salaries. It can be assumed that this is to offset the poor pay received by the crew from their employers, but when the employers also arrive to take a share, there are clearly other motives.

This offshore loss of income and effectively 'black market' activities inevitably contribute to a lack of investment in facilities for the resident population, and leads to further resentment (Villacis and Carillo 2013) (Fig. 4.8).

Although it can be said that the marine reserve has been relatively successful in preventing large industrial fishing boats from overfishing, its success in dealing with overfishing by the local fishermen is less clear. The sea-cucumber fishery which once generated much of the income of the local fishermen collapsed due to over-fishing, reflecting some of the problems associated with management of the marine reserve. Lobster fisheries, however, appear to now be somewhat more sustainable, but there may be an illusion of successful sustainable management due to natural fluctuations in the ecological baseline (Burbano et al. 2014; Castrejón et al. 2014).

In 2016, due to a large extent to political considerations, a new zoning and further restrictions were imposed within the Galapagos Marine Reserve, in the name of conservation, by the government of Ecuador. With the backing of the Charles Darwin Foundation and international organizations such as National Geographic, a new marine sanctuary was created to protect sharks across large areas of the oceans around the Galapagos. However, as the creation of the marine sanctuary was a top down initiative, many feel that in the long-run, it will not be able to increase the level of protection it is intended to. Inevitably, many fishermen consider that the new sanctuary is a threat to their livelihoods, but interestingly, it has not been well accepted either amongst many scientists and conservationists. One of the main criticisms is that it did not really increase the protection of the GMR as sharks were already protected everywhere in the reserve. It has also been criticized because agreement with some fishing groups was reached only after they were permitted, in a pilot study, to use a long line. Such lines had previously been made illegal in the marine reserve because of their proven

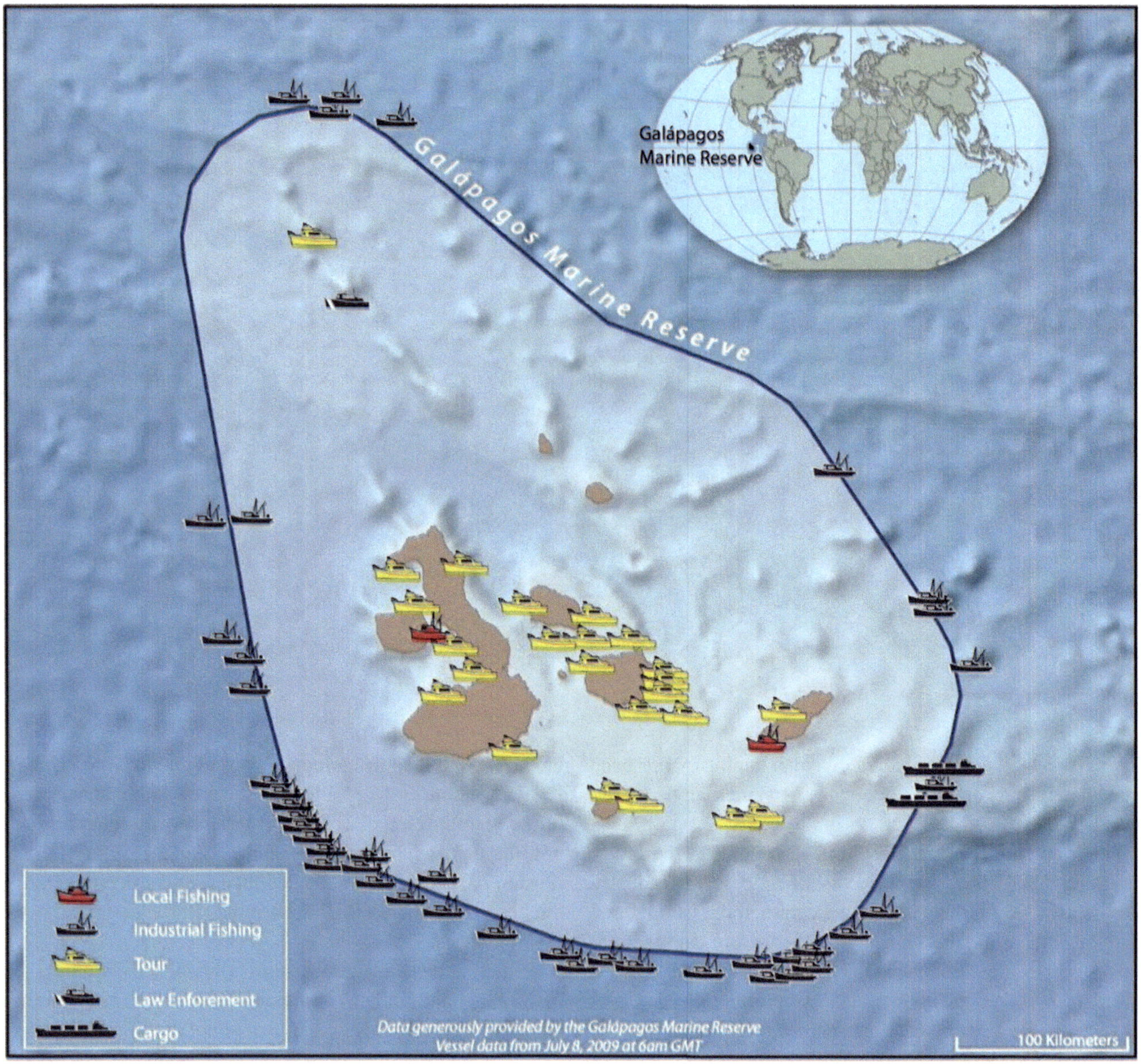

Fig. 4.8 The distribution of fishing across the Galapagos and the boundary of the Galapagos Marine Reserve (Map courtesy of Galápagos National Park)

impact on sharks, turtles and other important marine animals. Inevitably, there are also issues of non-enforcement, even corruption which means that even when regulations are in place, they can be abused or not always enforced (Quiroga 2009).

4.2.5 The Development of a Tourism Industry in the Galapagos

As tourism in Galapagos started to develop from the 1960s, it became seen as an essential part of the management of the Galapagos. The archipelago is ideally situated for tourism geographically, geologically and biologically, and creates an environment that now attracts hundreds of thousands of tourists each year. International and national managers, including within the Charles Darwin Foundation and the Galapagos National Park, argued that tourism could become a source of income for the both the Foundation and the National Park. There was, however, also a concern that increased numbers of tourists arriving to the islands could jeopardize the unique ecosystems present. Initially, it was proposed that international tourists would arrive in the Galapagos and stay in large cruise ships, so that they would not create any more demand on the island's limited natural resources. To support this model, the Charles Darwin

Research Station assisted in the recruitment of biologists and naturalists from Europe, North America and mainland Ecuador with appropriate scientific and language skills to become park guides (Reck 2017). Later, in the 1970s, funded courses for guides were developed by the Research Station, effectively subsidizing both the National Park and commercial tourism companies, who benefited from well-trained guides that were also acting as park rangers.

From the beginning there was the belief that too much tourism could threaten the Galapagos' ecosystems and biodiversity and initially, it was considered that 12,000 tourists per year was the carrying capacity. The number of tourist kept increasing and soon conservationists stopped trying to set a maximum limit for the number of tourists that can visit the island (Reck 2017). Nevertheless, although the total number of boats in the Galapagos has been regulated, the number of tourist has not been capped. In 2017 some 240,000 tourists arrived in the Galapagos, most coming from developed western countries, but increasingly Ecuadorians are beginning to constitute a larger percentage of these tourists (although the costs of travelling to the Galapagos are still beyond the reach of the majority of the national population). In the 1970s and 1980s, Santa Cruz became the headquarters for tourism operations, as well as for the Charles Darwin Research Foundation and the Galapagos National Park, hence stimulating the growth of the local population.

Although, the total number of cruise boats has not increased, there has been an increase in the number of people that can be lodge in the island's three main towns. In addition, although there were originally only 180 places that tourists could visit, new places were opened as pressure on the sites, including from the local population, increased. Cruise-based tourism and public and private institutions also led to the arrival of more residents. As a result, a positive feedback system has been established, and as more people visited the Galapagos, it became clear that more money was needed to improve the services and infrastructure of the Islands. Twenty-four hours a day electricity, better communication, faster transportation and better hotels and restaurants made the islands more attractive to tourists from around the world. The types of tourists arriving has also become more diverse, as different age and interest groups started to arrive. Young backpackers, national tourists, college and high school groups and people looking for adventure activities are amongst the new types of tourists which now mingle with a more 'traditional' nature-watcher visitors (Fig. 4.9).

These diverse groups demand other types of services above and beyond those which supported the large cruise boats, and increasingly these have been provided by the local population. In the 1990s, land based-island hoping tourism developed with the increased availability of hotels, residencies and restaurants in the towns. New types of activities such as SCUBA diving, surfing, sport fishing, kayaking and academic programs organized by national and foreign schools, institutes and universities as well as 'voluntourism' emerged in the 1990s and early 2000s. Many of these new types of activities were based in the towns, initially on Santa Cruz, but increasingly on San Cristobal and Isabela. Tourism has thus become the dominant activity that has brought new economic and human resources to the Galapagos and it now constitutes more than 50% of the income of the local economy. As more people started to work in the growing tourism industry, traditional productive activities such as fishing and agriculture have become less attractive for the residents. And as the percentage of people moving away from agriculture and fishing to tourism, a different vision of the Galapagos has begun to emerge with the economy increasingly becoming a service economy.

Nevertheless, many still view the increase in tourism as a threat to the wildlife and habitats of the Galapagos National Park and have often proposed quotas, but without any clear scientific assessment of a potential carrying capacity (and 240,000 people over the course of a year is not a large number by global tourism standards). Unfortunately, some still promote mechanisms to deter the increase in tourism using financial tariffs, in particular through increasing the entrance fees for the National Park (Nicholson 1987), which were already at 110$ per person in 2016. The inevitable consequences of such an approach, however, would be to make visiting the Galapagos an even more exclusive experience, affordable only by a rich elite (something akin to the original concept of exclusivity and exclusion promoted when the Galapagos National Park when it was first established). Inevitably, such approaches may also 'back-fire', as the more affluent the tourist, the more materialistic and disconnected they are likely to be from the nature of the islands. Indeed, irreparable damage has already occurred for the 'benefit' of such visitors, as an Iguana nesting site became the site for a new luxury hotel (Bassett 2009).

The new resources and opportunities available from tourism, however, have been not only the basis for an important material transformation of the society but also for a cultural transformation. Tourism is in many ways more compatible with conservation and a sustainable future for the Islands, not only because it provides an income based on non-extractive activities and increases the awareness on the wonders of the Galapagos, but also because it is a source of funds to support the Galapagos National Park, the Charles Darwin Foundation and other conservation efforts. Together with public campaigns, land based tourism is generating an important change in the visions and values of the local population (Quiroga 2013) that is now more receptive to the messages of conservation. Crucially, the realization that the local population is essential to such developments and

Fig. 4.9 Modern tourists waiting for their boats on the pier at Puerto Ayora. *Photo* D. F. Kelley

changes has perhaps also changed some of the more entrenched attitudes within the Charles Darwin Foundation and hence influenced the management of the national park. Rather than being a threat to the unique natural ecosystems and species of the islands, working with local people can now be shown to be essential part of its sustainable management (Quiroga 2013).

Very significantly, the increasing number of tourists visiting the islands now includes many Ecuadorians. This is linked to the gradual improvement of the national economy, meaning that an Ecuadorian 'middle class' now has more access to travel, especially after the currency was dollarized in the year 2000. Many Ecuadorian school-children now also visit the islands as part of their school trips. As most of these national tourists primarily use facilities in the towns, most of the money that they spend goes directly to the land-based operations in the islands which is not the case with much of the tourism associated with the cruise boats. Indeed, as many have observed, a major flaw in the tourism-supporting-the-population-supporting-conservation model is that most of the money generated by the latter style of tourism—perhaps as much as 90% of the total tourism revenue—simply goes offshore, to mainland Ecuador or elsewhere, hence providing no local benefits (Fig. 4.10).

An important aspect of the growing number of tourists arriving on the islands is that management of the Galapagos is increasingly being carried on the basis of the needs and expectations of the new visitor. An emphasis on certain charismatic species at the scientific level, also reflects the economic concerns of the managers. Increasingly, therefore, the eradication of introduced species, repopulation with species raised in breeding centers in Santa Cruz or others islands and management programs to save certain endemic plants and animals have been directed at the most charismatic species due to their 'value' to a growing and powerful tourism industry. Thus the Galapagos is today being 'crafted' as a nature destination, but in the process there is a risk that objective, scientifically-informed decision making might be affected.

Fig. 4.10 The reality of tourism in the Galapagos is that very little of the income stays in the islands and benefits the resident communities: extract from a panel in the Centro de Interpretacion Gianni Arismend, Pte. Baquerizo Moreno, Isla San Cristobal—red indicates proportion of financial benefit going offshore and green that remaining in the Galapagos. Note that although the majority of port-based service income stays in the Galapagos, such services only represent 6% of the total income generated by tourism in the Galapagos. The caption below the panel read: "*Profits from tourism should benefit a local population custodian of its natural assets and properly trained to manage them. What makes a fair distribution of these benefits*?"

During the zoning of the Galapagos Islands as part of the process of establishing the National Park and the Marine Reserve, around 180 sites were designated for tourist visits. Places selected constitute some of the most valued spaces on the Galapagos. Some, such as Bartolome Island and Tagus Bay were selected because of their scenic beauty, but in other cases, their biological interest, including the number of endemic species present, has been the main consideration. Many of the coastal areas include snorkeling as a permitted activity and some are used exclusively for SCUBA diving, in particular the islands of Darwin and Wolf (Fig. 4.11).

With the recent development of land-based tourism more places have been opened up on the main islands of Isabela, Santa Cruz, San Cristóbal, but also on Floreana. In some places, fishermen are now allowed to take tourists to areas that were once restricted solely for registered tourism providers, or were not even open to visitors, such as Punta Pucuna in San Cristobal and Cabo Rosa (Los Túneles) in Isabela. These and other similar schemes have allowed local fishermen to be more involved in tourism and many are abandoning fishing as they discover this can be an easier and better paid alternative. Not surprisingly, however, there is increasing tension around these new places as fishermen and day tour operators compete for access. The decision as to which places can be accessible and to which group is taken by the Galapagos National Park—including following

Fig. 4.11 Tourist snorkeling and 'selfie' with turtle, Isla Isabela. *Photo* J. Bello-Page

studies by the Universidad San Francisco de Quito—and often involves an analysis of the impact that opening these new sites may have and the relative impacts that the different groups might have.

Nevertheless, tourism can still have negative effects, and one of the main being that the economy can develop too rapidly, hence leading to ecological damage where development controls, for instance, are not enforced and new species of animals, plants, insects, bacteria and viruses are accidentally (or deliberately) introduced. The development of tourism and other activities based in the towns, also attracts residents to these urban centers, meaning that local production of foods has decreased, creating a greater dependence on the imports from the mainland. With more boats and flights bringing people and goods there is also more risk of introducing invasive species to the islands which can negatively impact on endemic animals and plants. Wikelski et al. (2004), for example, following a study of endemic and native birds, concluded that there was a relationship between the time humans had been on an island and the level of species extinction, regardless of what were the activities of the local population.

As a result, the control and eradication of many introduced species has become an important conservation priority and although a considerable amount of time and resources have been spent carrying out this activity, not all projects have been successful and many have failed. The best and most publicized example is the eradication of pigs and goats from many islands. The latter project was considered to be successful from a conservationist perspective, but was criticized by the local people. Indeed, these expensive interventions have achieved some of their goals but have also resulted in the unexpected effects. For example, the elimination of the large invasive herbivores from some of the islands where they had replaced the tortoises, has resulted in the uncontrolled growth of the vegetation that is now dominated by invasive plants (Quiroga and Rivas 2016) (see Fig. 2.46).

Remarkably, such consequences which would appear obvious to conservation managers in many developed countries, seem to have been unforeseen in the Galapagos. In an attempt to address the new problem where there is no longer an extant tortoise population, such as on Pinta, the National Park authorities decided to bring sterilized tortoises from Española to repopulate the now goat free island, and hence control the vegetation (Quiroga and Rivas 2016). Nevertheless, once woody vegetation, in particular, becomes established, there is no possibility that even a sterilized tortoise population can squeeze between the stems to control it. In many other parts of the world, for instance across Europe, a considerable amount of conservation effort is also required to keep such scrub under control, and a preferred method is often to re-introduce grazing animals. Maybe regulated populations of sterilized goats, rather than tortoises, might actually be a more effective tool in parts of the Galapagos for controlling invasive vegetation and hence maintaining a tortoise-friendly habitat? In a similar vein, the concept of having parks-without-people which has come from some conservationist quarters has parallels with attempts to remove 'invasive species' seen as a threat to particular habitats and species, for instance when the Charles Darwin Research Station suggested that residents on Isabela should move to other islands (Reck 2017). Such proposals also echo the establishment of the first national parks in the USA, when the indigenous population was 'removed' (see Sect. 4.1.2).

In many cases, however, attempts at eradication of invasive species have failed to achieve their goals and this is particularly true with attempts to eradicate many invasive plant species. A well-known example are the attempts to eradicate the hill raspberry (*Rubus niveus* Thunb.)—a native of Asia—which despite a significant input of economic and human resources, failed to produce any result. Such failures have caused some scientists and conservationists to consider new ways of dealing with invasive plants that will not waste scarce resources in vain attempts to achieve complete eradication, through allowing the indigenous ecosystem to 're-adjust' to the invader. Termed 'novel' or 'hybrid ecosystems' these ideas have opened up an interesting debate, not only about conservation and habitat restoration, but also about the very meaning of so-called pristine environments around the globe (Quiroga and Rivas 2016).

Although much of the research carried out in the Galapagos has inevitably been about evolutionary process (see Santander et al. 2009) until very recently, there has been a distinct lack of studies on social, cultural and economic problems (in part reflecting the perspective presented by many documentaries, tour companies and conservation agencies that there are no significant resident population on the Islands …). As in the case of invasive plants, it has often been implied that although resident, this population somehow does not belong to the Galapagos and hence is invasive and dangerous for the local ecology. This discourse has shaped the way in which local people are treated, and as in other parts of the world, the view that a National Park should be free of people, as established in the USA in the 19th century, is also present in the Galapagos and implies that no residents should live on the islands.

The creation of tensions between the top-down versus bottom-up approach to protected areas and especially within the management of the National Park creation, has been a trend since the implementation of its first annual strategic plan in late 20th Century. In this management of land and marine areas there is clear evidence of a belief that solutions cannot emerge from the local population. Although some practices implemented by the National Park, such as fishing quotas for spiny lobsters and sea-cucumbers, were established with some consultation with residents, in particular the Galapagos fishermen, others are often imposed on the local population after construction on the mainland—or elsewhere—and then imposed on the local population. Not surprisingly, the latter may reject such changes and refuse to participate. Attitudes are beginning to change, however, and at last the needs of island communities are beginning to be recognized, even by 'Western' commentators and scientists (for instance in Walsh and Mena 2013) but still much needs to improve to assure the emergence of bottom up strategies so that the local people can developed sustainable strategies for the management of their own resources.

4.2.6 Concluding Remarks

To understand the development of conservation policy and practice in the Galapagos it is essential to explore the way in which a 'western' colonist's concept of nature and nature conservation emerged and was later exported to other areas, especially where European and North American countries had influence. As discussed, the expansion of this concept of nature conservation constituted not only the expansion of a New World view, but also of practices that all too commonly excluded existing inhabitants, in a way which some have compared to 'ethnic cleansing'. The view that the new 'National Parks' somehow represented 'pristine' natural areas untouched by humans is a construct of this view, that rarely has any basis in reality.

The use of fire to 'manage' forest areas across most continents, including North America, followed by farming, for instance across the Amazon, Africa and Asia, together with increasingly sophisticated hunting techniques, have transformed 'natural' habitats across the planet (although some might indeed argue that human interaction is just another facet of 'nature'). That such areas might later be considered to be 'pristine', very much reflects a European perspective, where 1000s of years have human interaction with the landscape have extensively modified, even destroyed, many natural habitats and species through farming, forestry and industrial and urban development. In Europe and elsewhere, this wholesale change is reflected in the concept of 'semi-natural' habitats in conservation practice (see Fig. 4.4).

A realization of the extent of these changes, and a perception of 'loss', evolved in the 19th century from a fear of 'wild' places to a romanticism about 'wilderness'. It is easy to see, therefore, how a myth of wilderness then becomes a justification for establishing protected areas and part of a management process.

In the Galapagos there is no evidence of successful human colonization before the arrival of the Spanish in 1536, although some traces of pre-Columbian artefacts have indeed been recorded (Anderson et al. 2016). This lack of a long-established, even 'indigenous' human presence before 1536, forms part of the argument that all human activity across the islands is to a large extent disruptive—and hence in the Galapagos it is easy for scientists and conservationists to portray humans are as invasive and destructive. This perspective is further justified by the illusion that the Galapagos, 300 years after discovery by a lost Spanish ship—but not colonized until the 1830s—was still the untouched natural place where Darwin first formulated his ideas on natural selection. This narrative and the view that the islands represented somehow an untouched natural laboratory in which to study natural selection became institutionalized in the mid-20th century when the Galapagos National Park and the Charles Darwin Research Station were established. The reality of course, was that Darwin visited an area in which the wildlife was already being exploited on an industrial scale to provision ships from many countries.

Borrowing ideas from different parts of the world and especially from a European-influenced 'developed' world, a National Park was created with the vision—even dogma—that people should not be there. This vision, however, if not only fundamentally flawed, it has also backfired in many ways. A key theme in this flawed belief is the idea that the Galapagos constitutes a pristine and harmonic system that can be managed with a command and control strategy. As managers have tried to implement this concept, the dynamic and unpredictable character of nature gains the upper hand—for instance invasive species have taken hold, certainly irreversibly in many cases, and as soon as one is perceived to have been eliminated, a new ecosystem emerges that is different from the original one. That such things are inevitable is now, however, part of a general global culture as perfectly stated by the fictional scientist, Dr. Ian Malcolm, in the book (and later motion picture), Jurassic Park: "Life will find a way"—as the supposedly sterilized, genetically-resurrected dinosaurs run amok (Crichton 1990).

The other key flaw is the idea that the Galapagos can be managed without the participation of its residents, especially its historical inhabitants. This concept portrays nearly 190 years of interaction between residents and wildlife on the islands as entirely negative, it shows little concern for the future of such a population, should the natural resources it needs to survive be taken away. Various schemes have been developed over the years to further exclude the population from its own lands and livelihood, and even effectively prevent its own economic and social development—such as the 1960s idea of limiting tourism to externally controlled cruise ships and controlling visits to inhabited areas to prevent local development. Most of the management strategies that excluded participation of local people have now backfired and large dynamic towns have sprung-up, creating their own land-based tourism operations which in turn has promoted further development. Many local residents have developed interesting enterprises to provide services to the growing number of tourists, and many of these businesses could form the basis for a sustainable system of managing tourism in the Islands.

Thus the concept that the National Park can be controlled in a linear and predictable top-down manner, and can be managed and shaped to recreate a perceived, original 'pristine' state—and that it is better managed without local communities—has proven to be not only difficult to apply but also, ultimately, unproductive. In the Galapagos this local population has been, to a large extent, marginalized from most of the conservation practices of the National Park until very recently. This has generated a distance and tension between the Park's managers and the residents that is often difficult to solve and could take many years of dialogue to resolve. Nevertheless, the potential is there for a very positive future for the Galapagos, with a healthy, involved and educated resident population contributing to the successful and sustainable future for the National Park. Indeed, as pointed out by McCleary (2013), by working with the resident population, especially farmers, the GNP could find great allies in the fight against many invasive species, which can be as much a problem for the former as to the aims of the latter.

4.3 Geotourism, Geoparks and the Sustainable Economic Development in the Galapagos

4.3.1 Introduction

As can be seen, the development of conservation policy and practice in the Galapagos has had long and often conflictive origins due to the exclusion of local communities. Although things are now beginning to change, if these communities are to thrive, they will need to develop new activities and facilities—both products and services—both in collaboration with Galapagos National Park, but also within the areas they are permitted to inhabit and manage themselves, i.e. outside the 'boundaries' (in the widest sense) of the GNP. The perceived sensitivity of the endemic species of the islands to a human-presence, which is often used as a justification to exclude or drastically control human activity, is often misleading—sea lions sleeping on park benches and marine iguanas dozing around seaside bars in Santa Cruz are good examples. Elsewhere, Darwin's finches forage around cafés, Mocking Birds visit urban streets, frigate birds follow cruise ships and tortoises tolerate inquisitive tourists. Even Darwin noticed how 'tame' and apparently unconcerned by a human presence much of the wildlife of the islands was and still is—perhaps 500 years is not enough time for a wild species to evolve an innate fear of humans? (Figs. 4.12 and 4.13).

With the end of hunting of terrestrial wildlife, the main conservation issues for the archipelagos iconic wildlife revolve around invasive species, and who better placed to help with this un-ending battle, but a resident population? Crucially, this population also has the potential to help develop new nature experiences for island visitors, to help not only deflect increased pressure on key visitor sites *within* the National Park, but also to develop new, and ecologically less sensitive areas, and again spread tourists around. And the focus for these new experiences can be the rich and dramatic geodiversity of the islands, which is robust, omnipresent and spectacular.

4.3.2 Geotourism—An Introduction

The recognition of the potential of strange and spectacular geological features to attract visitors is nothing new. From the mystical associations, indigenous societies often have with the landscapes they inhabit—which is still as much true for some 'developed' regions as it is for those where more

Fig. 4.12 An endemic Galapagos Mockingbird (*Nesomimus parvulus*) on Playa Espumilla, Isla Santiago, showing a characteristic lack of fear of humans (and a disregard for the 2 m, tourist-wildlife rule). *Photo* J. Bello-Page

Fig. 4.13 Adaptation to human settlement and facilities—**a** a sea lion asleep on a park bench in Puerto Villamil, Isla Isabela. The inscription on the bench reads: "Isabela grows for you". *Photo* J. Bello-Page; **b** a marine iguana and a young tourist sizing each other up on the beach at Puerto Villamil, Isla Isabela. *Photo* D. F. Kelley

traditional ways of life still survive—to the fashionable aesthetics of 'sophisticated' travelers from the 18th century onwards, local 'tourism providers' have always benefited. At one level, the simple provision of supplies, and maybe accommodation, for visitors benefited a local economy—but in many cases, some form of paid guide would also be required (perhaps in addition to any 'entrance fee', such as a 'votive offering'). And then there are the souvenirs …

Early European examples of sophisticated tourist provision focused on geological, rather than purely cultural, features are discussed by Hose (2008). In the case of the Falls of Clyde and Staffa in Scotland (Gordon 2012), the attraction was not the remarkable fluvial geomorphology, or the drainage evolution of the region, it was the aesthetic beauty and dramatic natural power of the feature, as promoted by various contemporary artists. Similarly, the basalt columns of the Giants Causeway in the north of Ireland, attracted visitors due to their spectacular and bizarre structure, not as volcanologists or petrologists (Doughty 2008).

And this is where the many definitions of 'geotourism' which have been proposed—as mentioned in Chapter—start to become problematic. With the exception of a relatively small 'elite' of professionally qualified geoscientists, it would be entirely wrong to assume that most other visitors to geological features are inspired by such things as tectonic processes and magma cooling curves—the reality is that beauty and bizarre are still major motivations. But for anyone who has taken a University student group, or even the delegates of an international geological congress, to the more spectacular sites, the exclamations of awe and wonder that often expressed, reveal that even the most serious of geoscientists are not actually so fundamentally different from a more general public.

However, once inspired, the geological and conservation communities have an invaluable opportunity to, gently, educate the more general visitors about science and conservation, and why it matters to their society. One master's student group, in particular, from Birkbeck College of the University of London (UK) as part of Module on Earth Heritage Conservation in the 1990s, coined some 'new' terms for this process, firstly the visitor must be '*geoinspired*' and then they can become '*geoconnected*' to this heritage …

4.3.3 Geotourism and Geoheritage

Tourism is a cultural and economic process, and as the tourist derives pleasure and satisfaction through the experience of visiting a novel location away from their normal place of residence, so the residents of that location and facilitating 'third-party' providers can derive economic benefit from the visitors. In an ideal socio-economic context, this process can be symbiotic, but all too commonly, some party is exploited by another, with visitors sometimes exploiting the residents, or vice versa, but in many developing areas, some 'third-party' service providers may well be exploiting everyone.

In order to highlight the contribution of geological heritage features to tourism, in particular geomorphological such as landscapes, mountains, natural coastlines and karstic features, plus some more geological such as spectacular fossil localities and mines, the term 'geotourism' has become popular. There are many claims as to whom used the term first, but as the potential to raise the profile of the Earth Sciences, including to help fund research activity and employment opportunities, has been realized, a plethora of increasingly complex and restrictive definitions have been offered, including attempts to establish the subject as an 'applied science'. Some of these definitions are offered below, and although mainly from English language publications, some are commonly quoted in other national works (e.g. in Brilha 2016; Carcavilla et al. 2007, etc.). According to Hose (2006, p. 221), however, the first widely available use of the term is in Jenkins (1992) in a volume on Special Interest Tourism, but according to Hose, *Geotourism* is:

"*The provision of interpretive and service facilities to enable tourist to acquire knowledge and understanding of the geology and geomorphology of a site (including its contribution to the development of the Earth sciences) beyond the level of mere aesthetic appreciation*". (Hose 1995)

… which evolved to: "*The provision of interpretative facilities and services to promote the value and societal benefit of geological and geomorphological sites and their materials, and to ensure their conservation, for the use of students, tourists and other recreationalists*". (Hose 2000)

Stueve et al.'s definition of 2002 invokes inherent benefits to a local population, by defining Geotourism as: "*Tourism that sustains or enhances the geographic character of the place being visited – its environment, culture, aesthetics, heritage and well-being of its residents*".

According to Newsome and Dowling (2006), this is 'geographical tourism', and the authors provide their own very complete definition: "… *in our definition of geotourism the 'geo' part pertains to geology and geomorphology and the natural resources of landscape, landforms, fossil beds, rocks and minerals, with an emphasis on appreciating the processes that are creating and created such features, At the same time the tourism component ... involves visitation to geosites for the purposes of passive recreation, engaging a sense of wonder, appreciation and learning. In association ... there may be regular tours, specific activities and even the development of accommodation facilities. In addition there may be various forms of geosite management in place. We thus posit that geotourism is a distinct subsector of natural area tourism, and not a form of tourism that also includes wider cultural and heritage components or tourism that focuses on wildlife, all of which can be considered as distinct and separate aspects of tourism in their own right*". (Newsome and Dowling 2006).

This was later simplified to: "*A form of natural area tourism that specifically focusses on geology and landscape. It promotes tourism to geosites and the conservation of geodiversity and an understanding of Earth sciences through appreciation and learning. This is achieved through independent visits to geological features, use of geo-trails and view points, guided tours, geo-activities and patronage of geosite visitor centres*". (Newsome and Dowling 2010)

Pforr and Megerle (2006) also invoked themes of sustainability: "*… geotourism is..more than just geological tourism (e.g. 'fossicking'* [i.e. collecting geological specimens]*) or visiting caves, and embraces the identification of geo-objects, landscape marketing and interpretation of the geological heritage of a region in a sustainable manner.*" … and similarly, Gray (2008) invokes a link to geoconservation activities, defining Geotourism as: "*Tourism based on an area's geological or geomorphological resources that attempts to minimise the impacts of this tourism through geoconservation measures*" (Gray 2008) … and there are many other definitions …

A common issue with such 'definitions' is that they tend to confuse the actual or potential touristic attraction of geological features with the aspirations of the author or authors, in particular in relation to the sustainable and educative use of the resource. The deliberate exclusion of any reference to specimen-collecting activities, or 'fossicking'—which can be problematic for conservation—from any definition is particularly revealing. With the exception of a very small number of dedicated 'geo-enthusiasts' (both professional scientists and amateur enthusiasts), however, who may travel widely to experience specifically Earth Heritage features, most so-called 'geotourists' will be attracted by a range of natural and cultural features, also including wildlife and the historical buildings. Inevitably, trying to separate the specific contribution of geological heritage to their experience can become increasingly futile, but a realization that the tourist's interests are much more 'holistic' is crucial to the development of effective provision —including as part of any sustainable development strategy. Such integrated approaches are the essence of UNESCO Global Geoparks, as discussed below, a designation which recognizes the combination of an exceptional geological heritage with a comprehensive educational and sustainable development strategy.

Returning to the many definitions of 'geotourism', however, it is clear that the nature and attraction of the resource to the actual or potential visitor should be clearly separated from the way in which it is used. The latter will be subject to the interests and whims of the visitors, and could range from actively removing geological specimens to a more passive enjoyment without any notable contribution to their personal education or environmental awareness. It is for the site manager and responsible local and national authorities, as well as the providers or proponents of any activity to define any additional contexts, including any aspiration for education and raising public awareness of both geological and broader environmental issues. With the latter considerations in place, the essence of most existing definitions of 'geotourism' are perhaps more accurately expressed as aspirations for a 'sustainable geoeducational tourism'. The reality, however, will be that most 'geotourism' is simply a 'fuzzy-edged' facet of a much broader sphere of sustainable ecological and cultural tourism, rather than a wholly independent theme in its own right. In this context, the definition of Robinson (1998) is probably the most comprehensive, and geotourism becomes simply (Fig. 4.14):

"*Tourism in geological landscapes*". (Robinson 1998)

4.3.4 Geoheritage and Geoparks—A Model for the Sustainable Development of a Geoheritage Resource

The desire to use a natural heritage resource as a basis for the sustainable development of a local economy—where social and economic issues may be significant—is not new, and was perhaps first formally conceptualized within UNESCO's 'Man and the Biosphere' programme in 1971. 'MAB' is an intergovernmental scientific programme, launched in 1971 by UNESCO, that aims to establish a scientific basis for the improvement of relationships between people and their environments (https://en.wikipedia.org/wiki/Man_and_the_Biosphere_Programme). The concept led to the establishment of a global network of 'Biosphere Reserves', defined by UNESCO as:

"*Biosphere reserves are areas comprising terrestrial, marine and coastal ecosystems. Each reserve promotes solutions reconciling the conservation of biodiversity with its sustainable use.*

Biosphere reserves are 'Science for Sustainability support sites' – special places for testing interdisciplinary approaches to understanding and managing changes and interactions between social and ecological systems, including conflict prevention and management of biodiversity.

Fig. 4.14 Who are the real geotourists? Students on a Birkbeck College, University of London (UK) course on Earth Heritage Conservation observing students from the University of Liverpool studying geology (Saltern Cove Local Nature Reserve, English Riviera UNESCO Global Geopark, Torbay, SW England). *Photo* K. N. Page

Biosphere reserves are nominated by national governments and remain under the sovereign jurisdiction of the states where they are located. Their status is internationally recognized.

There are 686 biosphere reserves in 122 countries, including 20 transboundary sites."

(http://www.unesco.org/new/en/natural-sciences/environment/ecological-sciences/biosphere-reserves/).

The reserves are organised within the *World Network of Biosphere Reserves*, or WNBR, with the following vision:

"*The World Network of Biosphere Reserves of the MAB Programme consists of a dynamic and interactive network of sites of excellence. It fosters the harmonious integration of people and nature for sustainable development through participatory dialogue; knowledge sharing; poverty reduction and human well-being improvements; respect for cultural values and society's ability to cope with change. Accordingly, the WNBR is one of the main international tools to develop and implement sustainable development approaches in a wide array of contexts.*"

(http://www.unesco.org/new/en/natural-sciences/environment/ecological-sciences/biosphere-reserves/world-network-wnbr/).

Despite the criteria defined for the UNESCO programme, not all of the earlier 'Biosphere Reserves', as management entities, were able to fulfil these aims, and the limited regulatory mechanisms available to UNESCO inevitably led to some areas listed barely fulfilling any of any aims. Examples include the original North Devon Biosphere Reserve in SW England, declared in 1976, but largely focused on a single National Nature Reserve, only around 13.4 km^2, representing a coastal sand dune complex under private aristocratic ownership and dominated by a military training area—and consequently with virtually no resident population. Similarly, the original Odesa-Vignemale Biosphere reserve in the central Aragonese Pyrenees, Spain, was largely an area of uninhabited high mountain, including a USA-style National Park from which traditional use by the local community was largely blocked—so again with very little resident population and as in North Devon, hence virtually no local economy to sustainably develop. New initiatives in both areas have, however, have changed this scenario, and the North Devon Biosphere was expanded in 2002 expanded to around 3,300 km^2, including a broader river catchments areas with traditional agricultural practice (https://www.northdevonbiosphere.org.uk/; https://www.northdevonbiosphere.org.uk/uploads/1/5/4/4/15448192/north_devon_biosphere_periodic_review_2015.pdf) and the Odessa-Vignemale Biosphere Reserve is now included within the sustainable development strategies of the Sobrabre UNESCO Global Geopark (http://www.geoparquepirineos.com/contenidos.php?niv=1&cla=_2OA1CATPF&cla2=_2OA1CF1VM&cla3=&tip=2&idi=3).

Although it could be argued that both of these Biosphere Reserves have at their core geological and geomorphological systems and processes—coastal and fluvial systems for North Devon and a high Mountain belt of dominantly Upper Palaeozoic to Palaeogene age for Odessa-Vignemale,—the 'MAB' process emphasizes ecological features in relation to human health and well-being. Hence fundamentally geoheritage-focused areas could not become Biosphere Reserves. Although it would be entirely misleading to claim that there are no earlier examples of sites or areas with geological heritage features that have been developed for sustainable economic activities, the principles and practice that would much later evolve into the UNESCO Global Geoparks programme—the Geoheritage-equivalent of UNESCO Biosphere Reserves—were perhaps first developed in the 1980s within the French, Reserve Geologique de Haute Provence.

The Reserve was declared as part of a national network of natural heritage sites in France with a high level of protection in 1984, based on a national law of 1976 and now incorporates nearly 60 municipalities covering 200,000 ha (http://www.reserves-naturelles.org/geologique-de-haute-provence; www.resgeolo4.org; Gamet 1994). The area includes a wide range of geological and geomorphological features, but is scientifically best-known for its fossiliferous Jurassic and Cretaceous sequences, which includes globally important stratigraphical reference sections for a number of divisions of geological time.

By the late 1980s, the reserve included an interpretation center and several satellite sites with sign-boarding plus a range of educational activities and publications for the local population and visitors alike. In 1990, it was already reported that nearly 6,000 school pupils had been involved in these activities and the phrases such as "*Learn to read the Earth*" and "*Where the Memory of the Earth is protected*" adopted (Gamet 1994). It is not surprising, therefore, that when the Premier Symposium International sur la Protection du Patrimonie Geologique was held at Digne-les-Bains, at the core of the Reserve, in 1991, the famous mantra for subsequent geoconservation globally, the *International Declaration of the Rights of the Memory of the Earth,* the 'Digne Declaration', was penned (Martini and Pages 1994, Appendix; see http://www.progeo.ngo/downloads/DIGNE_DECLARATION.pdf):

"1. *Just as human life is recognized as being unique, the time has come to recognize the uniqueness of the Earth*"

"2. *Mother Earth supports us. We are each and all linked to her, she is the link between us.*"

"3. *The Earth is 4.5 billion years old and the cradle of life, of renewal and of the metamorphosis of life. Its long evolution, its slow rise to maturity, has shaped the environment in which we live.*"

"4. *Our history and the history of the Earth are closely linked. Its origins are our origins, its history is our history and its future will be our future.*"

"5. *The aspect of the Earth, its very being, is our environment. This environment is different, not only from that of the past, but also from that of the future. We are but the Earth's companion with no finality, we only pass by.*"

"6. *Just as an old tree keeps all the records of its growth and life, the Earth retains memories of its past ... A record inscribed both in its depths and on the surface, in the rocks and in the landscapes, a record which can be read and translated.*"

"7. *We have always been aware of the need to preserve our memories – i.e. our cultural heritage. Now the time has come to protect our natural heritage, the environment. The past of the Earth is no less important than that of human beings. Now it is time for us to learn to protect, and by doing so, learn about the past of the Earth, to read this book written before our advent: that is our geological heritage.*"

"8. *We and the Earth share our common heritage. We and governments are but the custodians of this heritage. Each and every human being should understand that the slightest depravation mutilates, destroys and leads to irreversible losses. Any form of development should respect the singularity of this heritage.*"

"9. *The participants of the 1st international symposium on the protection of our geological heritage, including over a hundred specialists from over thirty nations, urgently request all national and international authorities to take into consideration and to protect this heritage by means of all necessary legal, financial and organizational measures.*"

Within the Reserve, activities and presentations for a wider community also included local exhibitions, presentations to each municipality and collaborative projects, including documenting historical industries related to geological resources, as well as other cultural activities. Although it is clear from earlier publications that the emphasis of much of the initial activity was focused on raising awareness of geological heritage and hence engendering support for its conservation, the clear economic benefit of increasing visitor numbers to an area traditionally 'bypassed' from the main tourist routes from central and northern France to the Mediterranean became apparent. As a result, a concept of promoting economic development for the benefit of local communities became a key component of the development of the Reserve. By the mid-1990s, in collaboration with UNESCO's then Earth Science Division, this idea had evolved into a concept of Geoparks, as proposed at the 30th International Geological Congress in Beijing, China, in 1996 (Zouros and McKeever 2008). The aims of the new designation would include conserving geological heritage, improving public understanding of the Earth Sciences and promoting regional economic development (Patzak and Eder 1998; Eder 1999). Unfortunately, however, UNESCO did not adopt the proposals at the time, apparently due to political concerns about the implications of a new designation.

As already mentioned in Sect. 2.4.2, collaboration with three similar areas in Spain (Maestrazgo Cultural Park), Greece (Lesbos) and Germany (Vulkaneifel) led in 2000 to the founding of the European Geoparks Network (EGN) at a meeting in Molinos in Aragón, Spain (Teruel Province), within the Maestrazgo Cultural Park. To support the new network, funding from European Community funds for 'less favoured' areas, such as the 'Leeder II' programme would be sought to develop the necessary projects and economic infrastructure. Subsequent the four pioneering areas signed a charter and agreed a set of 'Operational Guidelines' in 2004, defining the essence of what a 'European Geopark' should be and how it should operate. This initial definition was as follows:

- "*A European Geopark is a territory which includes a particular geological heritage and a sustainable development strategy supported by a European programme to promote development.*"
- "*A European Geopark must comprise a certain number of geological sites of particular importance in terms of their scientific quality, rarity, aesthetic appeal or educational value*"
- "*A European Geopark has an active role in the economic development of its territory through raising the profile of geological heritage, including through the participation of its residents and promoting the development of geotourism. The Geopark should also support environmental education, training and research in aspects of the Earth Sciences and the enhancement of the local natural environment, including through sustainable development policies.*"

Submissions for approval of European Geoparks status were made by the local governmental or administrative body directly responsible for the management of the area proposed and assessed by a committee made up of representatives of established European Geoparks, which was followed a field evaluation. Despite increasingly complicated admission and evaluation procedures, by September 2007, 32 European Geoparks in 13 countries had been admitted to the Network.

Ironically, despite UNESCO's initial reluctance to adopt a formal Geoparks programme, the establishment of 11 national Geoparks by China in 2000 (Xun and Milly 2002) could not be ignored and soon a UNESCO-*recognised* network of Global Geoparks (GGN) came into existence in 2004. Note that this was still not a UNESCO *designation*, but simply an official 'recognition'. Not surprisingly, definitions of the designation were very similar to those for European Geoparks (www.unesco.org/science/earthsciences/geological_heritage):

- "*Geoparks are defined as sites or areas of geological significance, rarity or beauty, in which geological features play a significant part, and where the geological heritage is protected and developed at the same time.*"
- "*Geoparks shall foster socio-economic development that is culturally and environmentally sustainable by triggering tourism in the form of eco- or geo-tourism. Like this, a Geopark can have a considerable potential for economic development in rural and less developed regions.*"
- "*The management body of a Geopark shall take care of the logistic support for environmental education and training, research and monitoring, related to issues of conservation and sustainable development.*"

Following a memorandum of understanding, European Geoparks were automatically considered to be members of the new UNESCO-recognized Global Geoparks Network. By 2015, however, the success of the GGN had led UNESCO to formally recognize Global Geoparks as a formal designation, its first since the World Heritage Convention of 1972 (http://www.unesco.org/new/en/natural-sciences/environment/earth-sciences/unesco-global-geoparks) (Henriques and Brilha 2017).

To support this new initiative, the International Geoscience and Geoparks Programme (IGGP; http://www.unesco.org/new/en/natural-sciences/environment/earth-sciences/international-geoscience-and-geoparks-programme/) was established within UNESCO, incorporating two activities, the International Geoscience Programme (ICGP; http://www.unesco.org/new/en/natural-sciences/environment/earth-sciences/international-geoscience-programme/), a scientific co-operative project with the International Union of Geological Sciences (IUGS), and UNESCO Global Geoparks.

In this 'Brave New World':

"*UNESCO Global Geoparks, within the IGGP, are the mechanism of international cooperation by which areas of geological heritage of international value, through a bottom-up approach to conserving that heritage, support each other to engage with local communities to promote awareness of that heritage and adopt a sustainable approach to the development of the area.*" (Statutes of the IGGP, Part B (UNESCO Global Geoparks), Article 1: http://unesdoc.unesco.org/images/0026/002606/260675e.pdf)

This definition is further expanded under the UNESCO website FAQ, "*What is a UNESCO Global Geopark?*":

"*UNESCO Global Geoparks are single, unified geographical areas where sites and landscapes of international geological significance are managed with a holistic concept of protection, education and sustainable development. A UNESCO Global Geopark uses its geological heritage, in connection with all other aspects of the area's natural and cultural heritage, to enhance awareness and understanding of key issues facing society, such as using our earth's resources sustainably, mitigating the effects of climate change and reducing natural disasters-related risks. By raising awareness of the importance of the area's geological heritage in history and society today, UNESCO Global Geoparks give local people a sense of pride in their region and strengthen their identification with the area. The creation of innovative local enterprises, new jobs and high quality training courses is stimulated as new sources of revenue are generated through geotourism, while the geological resources of the area are protected.*"

"*UNESCO Global Geoparks empower local communities and give them the opportunity to develop cohesive partnerships with the common goal of promoting the area's significant geological processes, features, periods of time, historical themes linked to geology, or outstanding geological beauty. UNESCO Global Geoparks are established through a bottom-up process involving all relevant local and regional stakeholders and authorities in the area (e.g. land owners, community groups, tourism providers, indigenous people, and local organizations). This process requires firm commitment by the local communities, a strong local multiple partnership with long-term public and political support, and the development of a comprehensive strategy that will meet all of the communities' goals while showcasing and protecting the area's geological heritage.*"

(http://www.unesco.org/new/en/natural-sciences/environment/earth-sciences/unesco-global-geoparks/frequently-asked-questions/what-is-a-unesco-global-geopark/)

Several other phrases are of particular relevance in the context of the Galapagos:

- "*A UNESCO Global Geopark must contain geology of international significance.*"
- "*UNESCO Global Geoparks are living, working landscapes where science and local communities engage in a mutually beneficial way.*"
- "*Education at all levels is at the core of the UNESCO Global Geopark concept. From university researchers to local community groups, UNESCO Global Geoparks encourage awareness of the story of the planet as read in the rocks, landscape and ongoing geological processes. UNESCO Global Geoparks also promote the links between geological heritage and all other aspects of the area's natural and cultural heritage, clearly demonstrating that geodiversity is the foundation of all ecosystems and the basis of human interaction with the landscape.*"

('Operational Guidelines' at: http://www.unesco.org/new/fileadmin/MULTIMEDIA/HQ/SC/pdf/IGGP_UGG_Statutes_Guidelines_EN.pdf).

Although now a prestigious international designation, with around 140 global members by 2018 (Henriques and Brilha 2017), the application process for becoming a UNESCO Global Geopark status has become very complex and expensive, and in some cases certainly not a genuine 'bottom-up' approach led by local communities as UNESCO would like observers to believe. Indeed, in some states with limited democratic freedoms when compared to European, one has to query whether a truly 'bottom-up' approach is really possible. In practice, most applications are led by the local or regional government authorities with administrative authority over the 'Aspiring Geopark' area rather than by any community group. This is not surprising, however, due to the very significant costs implicit in the application process, including the compilation of a comprehensive dossier for the area, with a full justification of international scientific importance, and strategies for management and education use. Applicant areas must also now demonstrate that such strategies are already in place at the time of application ('*Guidelines*' paragraphs 4, 5).

Additional costs include financing initial evaluation and, every four years, 're-evaluation' by assessors appointed from the global network (*Guidelines* paragraph 5.3). This is in addition to a compulsory annual 'contribution' of 'at least' $1,000 USD to UNESCO (*Guidelines*, paragraph 6) and an expectation that representatives of each Global Geopark will attend international meetings both annually and bi-annually. Only in "*exceptional circumstances*" and for "*developing countries only*" can such costs be waived or financial support sought from the UNESCO UGG secretariat.

Where local and governmental 'seed money', for instance for economic regeneration, is available, this process can work very well—the great success of the Global Geoparks initiative in some countries being very evident. However, in poorer regions and those with less supportive or problematic administrative regimes, despite genuine community support, such an approach can be daunting, or simply impossible (it is not surprising, therefore, that across Africa and most of central and South America there are few UNESCO Global Geoparks …). Even where grants might be available to initiate the process, the ongoing, inherent costs of continuing to belong to the Network may simply be unsustainable for some communities.

In a somewhat analogous manner in which the 'western' model of National Parks has been exported, the issue here is arguably the Eurocentric development of the Geoparks concept as already observed by Sá et al. (2017). Within Europe, for instance, there is often access to European Union support for sustainable-development projects in poorer areas, such as the original 'Leeder' funding that made the initial EGN project possible. Such funds, in combination with the development of low-cost 'budget' airlines, make regular Network meetings possible (although with the increasing size of the network and the necessity for larger venues have contributed to such meetings becoming increasingly expensive). In addition, the new requirement that the assessment of the international scientific importance of the area should be "*… based on the international peer-reviewed, published research conducted on the geological sites within the area …*" (*Guidelines* paragraph 2.2) creates other issues. This requirement may work well within Europe or the USA, where many international scientific publishers are based, but ignores the difficulties many in other parts of the world may experience in accessing such literature due to issues of both national scientific funding and expertise—as well as language, as such journals are now always published in English.

Crucially, the European concept of promoting education and community participation does not always reflect the preoccupations of communities in developing areas, where scientific studies may be few and there will be more immediate concerns about food and security. Sá et al. (2017), in particular, for Latin America countries (or 'LACs'), consider that "*… due to the specific social, political and cultural realities, the* ["*European Geopark Model*"] *is inappropriate taking into account the existing evaluation criteria of the UGG Evaluation Form.*" They also note that "*… the systemic lack of understanding in LAC countries about the fundamental principles of the UGG, despite their widespread outreach, has resulted in not so good practices and also in a wrong use of the Geopark concept as designated by UNESCO, creating in many cases a negative social interpretation about the UGG territories. This reality in the LAC countries, contributes to an abusive process of new consultants in this field so-called experts in the guidelines of the International Geoscience and Geoparks Programme, who only seek for a quick and easy profit and jeopardizing the proper development of new UNESCO Global Geoparks, as territorial development strategy made with people and for the people included in these territories.*".

They conclude that "*The LAC framework demands the development of new ways of thinking and approaches as well as new tools and strategies …*"

Concerns about other aspects of the direction that parts of the UNESCO Global Geoparks movement has taken have also been raised within Europe, and authors such as Hose and Vasiljevic (2012) note that there is growing concern within the established European geoconservation community about the approach and many of the activities of Geoparks. In particular, they note that proposals and management plans for Geoparks can often have limited consideration of geoconservation and geo-education, and all too commonly almost no attention is given to supporting the geological research that underpins these activities. Hose and Vasiljevic also observe that the increasingly insular operations of the UGG Network has begun to generate an increasingly 'parochial geopark literature' (p. 36). Although the latter's study focused on the principles and practice of Geotourism, the same problem also now seems to be developing for aspects of scientific assessments.

Nevertheless, with the establishment of the Latin America and Caribbean Geoparks Network (GEO-LAC) in 2018 and a cycle of Geopark meetings now taking place in Ecuador (as mentioned in Chap. 2)—as well as changes within UNESCO —there is now, hopefully, some possibility that the UGG statutes will now be able to change, and much better reflect the Latin-American context... An additional and unfortunate bi-product of the new UNESCO 'Global' Geoparks Network designation, however, is that the only countries recognized by UNESCO can be members (Guidelines, Paragraph 5.4)— hence the highly-successful group of previously 'Global Geoparks' in Taiwan, including Yehlui (http://www.ylgeopark.org.tw/ENG/info/YlIntroduction_en.aspx), were expelled from the GGN when 'Global Geoparks' became a formal UNESCO designation in 2015.

4.3.5 National and 'Local Geoparks'

Not surprisingly, given the complexity and implied costs of joining the UNESCO Global Geoparks Network, some areas have opted not to follow this route. Instead, they have established educational and economic regeneration projects and programmes for defined areas based on the original principles of the European Geoparks movement, but without any of the inherent 'external' obligations and costs. Crucially, administrative set-up and annual costs implied through applications to and membership of the GGN, can remain locally to fund employment and community activities. Crucially, despite frequent implications to the contrary, the use of the term 'Geopark' to describe such areas can be entirely appropriate and unchallengeable, as only terms such as 'European Geopark' and 'UNESCO Global Geopark'—and their related logos—have a legally protected status. Indeed, in order, to clarify this frequent 'misunderstanding', the IV International Symposium ProGEO on the Conservation of the Geological Heritage, held at the University of Minho in Portugal in September 2005—and convened by ProGEO, arguably the most influential global 'mover' in the field of Geoconservation,—agreed a declaration including the following statement: "*We endorse all local, national and wider development of Geoparks, which must be based on sound and sustainable protection of the geosite resource*" (http://www.progeo.ngo/downloads/DECLARATION_OF_BRAGA.pdf).

A number of countries such as Germany (www.nationaler-geopark.de), China (Yang et al. 2011) and Korea (http://www.koreageoparks.kr) already have well established national networks of Geoparks, framed within national guidelines to ensure that any included site or area fulfils certain basic criteria related to its Geoheritage significance, management infrastructure and educational and touristic provision. Crucially, countries such as these also have UNESCO Global Geoparks networks, proving that both systems can work in parallel with no confusion as to their relative international status. In other countries, such as the UK, other Geoparks also exist, either explicitly, or de facto in the way they are managed, although no agreed national guidelines currently exist. For some proponents of the UNESCO Global Geopark concept, this lack of 'regulation' may be problematic, but in reality, the actual contribution of the defined area to the community, both educationally and economically, can only be judged on a case by case benefit, as indeed for any UNESCO approved Global Geopark. This would be the "proof in the [Geopark] pudding".

A good example of such an independent initiative is the South Wales Coalfields Geopark (Wales, UK), led by the National Museum of Wales (NMW), in an area of high unemployment and deprivation as a result of the closing of the traditional deep coal mining industry (Evans 2005). In this area, activities are primarily led by a single Geopark officer, typically in partnership with various local and national organisations, including the NMW in its broader community-outreach role. The project has been very successful in obtaining grants from both local municipalities and sponsorship from local businesses and after the salary and administrative costs of the project and the project officer are covered, all other monies generated can go straight into the community to fund educational and other activities (Evans et al. 2017).

Within this wider sphere of sustainability-linked, community-based Geoheritage activities, one new concept stands out—that of *Geovillages*. This is truly a community-led, 'bottom-up' initiative, where activities are volunteer led and there is no employed executive and no external constraints on what can be developed and achieved. Although originated in the village of Martley in western England, links are now developing with other community-based activities elsewhere in Europe and there are ambitions to establish an informal global network (http://www.geo-village.eu/wp-content/uploads/2016/11/Final-version-of-MAZURSKI-paper.pdf). And 'informal' is the key concept, the founding idea is to celebrate a local geological heritage—and enjoy the process!

4.3.6 The Geopark Concept and Sustainable Development in the Galapagos

The European Geoparks concept, as exported as 'Global Geoparks', forms a very interesting model for any aspiration of *sustainable* develop based on a geological heritage (and the latter, as discussed in Chap.2, is what the Galapagos has in abundance). Crucially, significant and spectacular elements of this geodiversity, remain under the control of the resident communities as they lie outside of the borders of the National Park, but other key features, within the Park, are still used by the community in their role as registered guides or excursion providers. Hence, there is a very real potential to begin to develop touristic activities focused on such features which can genuinely contribute to the economic and social development of island communities. In addition, the integration of such activities within other existing and developing sustainable development initiatives such as rural visitor accommodation and the promotion of sustainable agricultural practice, provides just the structure which has proved so successful for UNESCO Geoparks globally.

Crucially the relative robustness and immovability of geological features means that an increased use for tourist visits can be more sustainable than more delicate ecological features. In addition, geological features are present throughout the islands, not only in National Park areas, so their use can even be developed independently of any of the constraints placed within the protected area (Fig. 4.15).

Although the issue of commercial-style 'branding' can often be a distraction within the UNESCO Global Geoparks process, in the case of the Galapagos, there can be little doubt that the area would tick most of the assessment boxes for global scientific value, and integration of a rich geoheritage with the iconic species and habitats. The current lack of

Fig. 4.15 Example of a robust geological feature in the Galapagos—the lava plain between Punta Moreno and Sierra Negra (in the distance), Isla Isabela. *Photo* J. Bello-Page

significant community involvement in the management of the GNP areas might be problematic, however, but with successful community based activities, a *de facto* Geopark could still be established, focussing on those areas which the resident community has influence over—but with the potential to, ultimately, extend its sustainable development practices across other areas. Although UNESCO may prefer not to have 'overlapping' designations, as stated in Global Geopark selection *Guidelines* criterion (iv): "*In the case where an applying area overlaps with another UNESCO designated site, such as a World Heritage Site or Biosphere Reserve, the request must be clearly justified and evidence must be provided for how UNESCO Global Geopark status will add value by being both independently branded and in synergy with the other designations.*"

But what is crucial, however, is the strength of the original European Geopark model, in that all sustainable development and community-focused activities, and groups promoting them, can be linked and integrated under a single banner and promoted with a clear identity and stated aims. Within a UNESCO Global Geoparks model, this would be a local governmental authority, but there are examples where the lead has come from community groups and voluntary associations, such as the Martley '*Geovillage*'. Such an approach has a lot of potential where local political or administrative issues might be complex and reaching agreement and promoting appropriate actions in a co-ordinated way might not be straightforward.

Inevitably, recognition in terms of a UNESCO Global Geopark designation could be a long-term goal, but the complex application procedures, implied costs and the need to have a promoting governmental authority, will put this beyond capacity of most community groups. Nevertheless, as in the case of the South Wales Coalfields Geopark, there may be no intention that a functioning Geopark area would go down the UNESCO route, as maximizing the direct community benefit maybe the primary aim. However, if a community led and genuinely sustainable initiative is successful in the Galapagos, it may even begin to influence some aspects of the management of the National Park and even encourage the Charles Darwin Research Center to join a partnership. Through such collaborations, it may well be possible to start to address some of the fundamental problems the communities of the Galapagos face, such as high levels of restriction, and the dominance of 'off-shore' companies in, for instance, touristic provision and fishing (and which contribute little or nothing to the economic and social development of the islands themselves).

Taking the essence of the original European Geoparks initiative—and its successful and innovative implementation across various existing Geoparks—combined with the most appropriate elements of its development as a Global program through UNESCO, has a great potential for providing a blue-print for sustainable development linked to a geological heritage elsewhere, including of course in the Galapagos. Currently, however, many examples of tourist activities exist across the inhabited parts of the islands, but remarkably few, have any significant Geoheritage content, even where the features visited are by definition geological, such as lava caves, volcanic craters and lava flows. Clearly there is considerable potential here to develop further activities, including thematic tours as a general tourist offer (rather than by special arrangement), as well as producing guides and other publications for visitors and developing thematic interpretative features and centers.

Indeed, models of how to sustainably develop the geological heritage of volcanic islands for the benefit of their resident populations exist elsewhere in the world, perhaps in the most comparable setting in the Azores archipelago in the central Atlantic, and now a UNESCO Global Geopark (Nunes 2014; https://www.azoresgeopark.com/?lang=EN). These islands represent a very similar suite of both active and extinct volcanic complexes to the Galapagos—in their case spanning the Mid Atlantic Ridge, an active plate tectonic spreading zone. Albeit in a very different and much wetter climactic regime—and hence with a much larger resident population—the range of activities focused on the regions rich volcanic heritage, could provide models for adaptation to the special circumstances of the Galapagos. The area also has its special problems with invasive species affecting its special island biota (Trota and Pereira 2015) but unlike the Galapagos, its Geological heritage has been inventoried and geotourism has been developed amongst its 121 protected Geosites (Lima et al. 2009, 2018). Linked activities include: Inter-island touristic circuits or routes, including both by boat and plane; Volcanic show caves (i.e. managed touristic access into lava tunnels); 'Belvederes', i.e. viewpoints for key geological features and landscapes, typically with installed site-information panels; Thematic Walking Trails, including coastal and urban Georoutes; Interpretation centers and museums, often combining cultural and ecological themes with geological; whale and dolphin watching, which often incorporates coastal geological views; hot springs and fumarolic activity, some of which may be available for bathing; guided tours and expedition; 'Geoproducts' including stamps and other souvenirs; and guidebooks, leaflets and web-based resources—including downloadable site based information—linked to a network of on-site marker posts (Fig. 4.16). Erfurt-Cooper (2014) provides accounts of a range of other developed, or developing, volcanic tourist destinations, including observations on the Galapagos by Dowling (2014), which provide additional useful information about other approaches.

To develop such diverse range of activities within a single community-focused vision for sustainable development clearly needs some form of administrative group to promote the concept and organize collective activities, plus set the principles and standards against which the activities of any prospective partners might be approved. As such there may be existing community groups and collective initiatives across the archipelago which could take, at least initially, a lead. Ideally, of course, support and collaboration with the Charles Darwin Research Centre and the Galapagos National Park, plus regional governmental organizations, could help strengthen and build any initiative. But such associations such as these can be a 'double-edged sword' and could even restrict or inhibit the development of a genuinely community-led initiative through political and other forms of control. Crucially, however, and as mentioned in Chap. 2 and above, there are now a number of Global Geopark projects developing within Ecuador from which to seek advice and guidance, even support. Successfully building on

Fig. 4.16 A classic view across the Sete Cidades caldera complex from an established view point or 'Belvedere'; Ilha Sao Miguel, Azores. *Photo* K. N. Page

community action can, however, also be very positive politically, and once achieved, collaboration with administrative authorities can come from a position of relative strength. Having achieved such a position, even possible recognition by UNESCO as a Global Geopark could be the next step, hence consolidating the role of Galapogean communities in the genuinely sustainable management of one of the World' most iconic natural regions.

References

Allen DE (1976) The naturalists in Britain: a social history Allen Lane

Anderson A, Stothert K, Martinsson-Wallin H, Wallin P, Flett I, Haberle S, Heijnis H, Rhodes E (2016) Reconsidering precolumbian human colonization in the Galápagos Islands, Republic of Ecuador. Latin Am Antiq 27(2):169–183

Anderson D, Grove R (eds) (1987) Conservation in Africa; people, policies and practice. Cambridge University Press, Cambridge, pp 355

Arnhart L (2007) Friedrich Hayek's Darwinian conservativism in liberalism conservativism, and Hayek's idea of idea of spontaneous order. Edited Louis Hunt and Peter McNamara. Palgrave Macmillan

Bassett CA (2009) Galapagos at the crossroads: pirates, biologists, tourists, and creationists battle for Darwin's cradle of evolution. National Geographic

Beinart W, Coates P (1995) Nature reserves and national parks. Environment and history: the taming of nature in the USA and South Africa, pp 72–92

Bonham-Carter V (1971) The survival of the English countryside. The survival of the English countryside, 240 pp

Botkin (2012) The Moon in the Nautilus Shell: discordant harmonies reconsidered. Oxford University Press

Brewington L (2013) The double bind of tourism in Galapagos society. In: Walsh SJ, Mena CF (eds) Science and conservation in the Galápagos Island: frameworks and perspectives. Springer, London, pp 105–126

Brilha J (2005) Patrimonio, geologico e geoconservacao. Palimage, Braga

Brilha J (2016) Inventory and quantitative assessment of geosites and geodiversity sites: a review. Geoheritage 8(2):119–134

Burbano D, Mena CF, Guarderas P, Vinueza L, Reck G (2014) Shifting baselines in the Galapagos white fin fishery, using fisher's anecdotes to reassess fisheries management: the case of the Galapagos grouper (Chapter 11). In: Denkinger J, Vinueza L (eds) The Galapagos marine reserve, social and ecological interactions in the Galapagos Islands. https://doi.org/10.1007/978-3-319-02769-2_11

Carcavilla L, López-Martínez J, Durán Valsero JJ (2007) Patrimonio geologic y geodiversidad: invewtigaci'on, conservación, gestión y

realción con los espacios naturales protegidos. Instituto Geológica y Minero de España, 360 pp
Carruthers J (2007) Influences on wildlife management and conservation biology in South Africa c. 1900 to c. 1940. South Afr Hist J 58 (1):65–90
Caruthers J (1989) Creating a national park, 1910 to 1926. J South Afr Stud 15(2):188216c
Castrejón M, Defeo O, Reck G, Charlese A (2014) Fishery science in Galapagos: from a resource-focused to a social–ecological systems approach (Chapter 8). In: Denkinger J, Vinueza L (eds) The Galapagos marine reserve, social and ecological interactions in the Galapagos Islands. https://doi.org/10.1007/978-3-319-02769-2_11
Chase A (1987) Playing god in yellowstone: the destruction of America's first national park. Harcourt Brace Jovanovich
Crichton M (1990) Jurassic Park. Alfred A, Knopf, New York, 400 pp
Cronon W (1995) The problem with wilderness in uncommon ground: toward reinventing nature (William Cronon, ed). W.W. Norton, New York
De Wever P, Alterio I, Egoroff G, Cornée A, Bobrowsky P, Collin G, Duranthon F, Hill W, Lalanne A, Page KN (2015) Geoheritage, a national inventory in France. Geoheritage 7:205–247
Desmond A, Moore J (1991) Darwin. Penguin, London, 808pp
Doughty P (2008) How things began: the origins of geological conservation. Geological Society, London, Special Publications 300 (1):7–16
Dowling R (2014) Galápagos: an emerging volcano tourism destination. In: Erfurt-Cooper (ed) Volcanic tourist destinations. Geoheritage, Geoparks and Geotourism, Springer, Berlin, pp 193–199
Eder W (1999) Geoparks of the future. Earth Herit 12:21
Erfurt-Cooper P (ed) (2014) Volcanic tourist destinations. Geoheritage, geoparks and geotourism series. Springer, Berlin, 384 pp
Evans D (1997) A history of nature conservation in Britain (2nd edn). Rutledge
Evans BG (2005) Geoparks, coalfields and South Wales a sustainable combination? Zeitschrift der Deutschen Gesellschaft für Geowissenschaften (ZDGG) Band 156 Heft 3 (2005):373–380
Evans BG, Cleal CJ, Thoma BA (2017) Geotourism in an industrial setting: the South Wales coalfield geoheritage network. Geoheritage. https://doi.org/10.1007/s12371-017-0226-3
Everhart WC (1983) The National Park Service. The National Park Service (2nd edn)
Folke C (2004) Traditional knowledge in social-ecological systems. Ecol Soc 9(3):7. [online] URL: http://www.ecologyandsociety.org/vol9/iss3/art7/
Gamet J (1994) Actions pedagogiques de la Reserve geologique de Haute-Provence. Connaiter pour aimer, aimer for proteger. In: Martini G, Pages JS (eds) Actes du Premier Symposium International sur la protection du Patrimonie Geologique, Digne-les-Bains, 11–16 juin, 1991. Mémoires de la Société Géologique de France, Nouvelle Série 165, pp 243–246
Gordon JE (2012) Rediscovering a sense of wonder: geoheritage, geotourism and cultural landscape experiences. In: Hose T (ed) Special issue: geotourism and geoconservation. Geoheritage 4(1–2):65–78
Gray M (2008) Geodiversity: the origin of a new paradigm. In Burek CV, Prosser CD (eds) The history of geoconservation. Geological Society Special Publication 300, pp 17–30
Griffiths T, Robin, L (1997) Ecology and empire: environmental history of settler societies. Edinburgh University Press
Harroy JP (1972) Addendum—corrigendum to the second edition of the United Nations lists of national parks and equivalent reserves
Hennessy E (2017) Mythologizing Darwin's Islands. Chapter 5 in Diego Quiroga and Ana Sevilla eds. Darwin, Darwinism and conservation in the Galapagos Islands. Springer
Henriques MH, Brilha J (2017) UNESCO global geoparks: a strategy towards global understanding and sustainability. Episodes 40:349–355
Hose TA (1995) Selling the story of Britain's stone. Environ Interpret 10(2):16–17
Hose TA. (2000) European geotourism—geological interpretation and geoconservation promotionfor tourists. Geological heritage: its conservation and management. Madrid
Hose TA (2006) Leading the field: a contextual analysis of the field-excursion and the field-guide in England. Proceedings of the critical issues in leisure and tourism education conference. High Wycombe
Hose TA (2008) Towards a history of geotourism: definitions, antecedents and the future. In: Burek CD, Prosser CD (eds) The history of geotourism. Geological Society of London Special Publication 300, pp 37–60
Hose TA (ed) (2016) Geoheritage and geotourism: a european perspective. The Boydell Press, Woodbridge
Hose TA, Vasiljevic DA (2012) Defining the nature and purpose of modern geotourism with particular reference to the United Kingdom and South-East Europe. Geoheritage 4:25–43
Howkins A, Orsi J, Fiege M (eds) (2016) National parks beyond the nation: global perspectives on "America's Best Idea" (vol 1). University of Oklahoma Press
Ise J (1961) Our national park policy: a critical history (No. SB482. A1 I75)
James MJ (2017) Collecting evolution: the galapagos expedition that vindicated darwin. Oxford University Press
Jenkins JM (1992) Fossikers and rockhounds in northern new south wales. In: Weiler B, Hall CM (eds) Special interest tourism. Belhaven, London, pp 129–140
Kantor I (2007) Ethnic cleansing and America's creation of National Parks. 28 Pub. Land & Resources L. Rev. 41
Kirby W (1825) Introductory address. Zool J 2:1–8
Lima EA, Nunes JC, Costa MP, Porteiro EA (2009) O geoturismo como instrumento de valorização do "Geoparque Açores". In: Carvalho CN, Rodrigues J, Jacinto A (eds) Geoturismo e desenvolvimento local. Câmara Municipal de Idanha-a-Nova/Geoparque NaturTejo, Lda., pp 149–160
Lima E, Machado M, Guerreiro M, Nunes JC, Costa MP (2018) Geological heritage management in small islands: the example of the Azores UNESCO Global Geopark (Portugal). Geoheritage (in press)
Lindsey A, Romanach S (2007) Economic and conservation significance of the trophy hunting industry in Sub-Saharan Africa. Biol Conserv 134(4):455–469
MacKenzie JM (1988) The empire of nature: hunting. Conservation and British Imperialism, Manchester, pp 43
Marks SA (1984) The imperial lion: human dimensions of wildlife management in Central Africa. Westview Pr
Martini G, Pages JS (eds) (1994) Actes du Premier Symposium International sur la protection du Patrimonie Geologique, Digne-les-Bains, 11–16 juin, 1991. Mémoires de la Société Géologique de France, Nouvelle Série 165
McCleary AL (2013) Characterizing contemporary land use/cover change on Isabella Island, Galápagos. In: Walsh SJ, Mena CF (eds) Science and conservation in the Galápagos Island: frameworks and perspectives. Springer, London, pp 155–172
Nelson JG, Needham RD, Mann DL (eds) (1978) International experience with national parks and related reserves, Department of Geography, University of Waterloo, Waterloo, pp 1–27
Neumann R (1998) Imposing wilderness: struggles over livelihoods and nature preservation in Africa
Newsome D, Dowling RK (2006) The scope and nature of Geotourism. In: Dowling RK, Newsome D (eds) Geotourism. Elsevier/Heineman Publishers, Oxford, pp 3–25
Newsome D, Dowling RK (eds) (2010) Geotourism: the tourism of geology and landscape. Goodfellow Publishers, Oxford

Nicholson EM (1970) The environmental revolution. Hodder and Stoughton

Nicholson EM (1987) The new environmental age. Cambridge University Press, Cambridge

Nunes JC (2014) The azores archipelago: islands of geodiversity. In: Erfurt-Cooper (ed) Volcanic tourist destinations. Geoheritage, Geoparks and Geotourism, Springer, Berlin, pp 57–67

Page R, Bentley M, Waldrop J (2013) People live here: maternal and child health on Isla Isabela, Galapagos. In: Walsh SJ, Mena CF (eds) Science and conservation in the Galápagos Island: frameworks and perspectives. Springer, London, pp 141–145

Patzak M, Eder W (1998) "UNESCO Geopark". A new programme—a new UNESCO label. Geologica Balcanica 28:33–35

Pforr C, Megerle A (2006) Geotourism: a perspective from southwest Germany. Geotourism, pp 118–139

Pritchard PCH (1996) The Galapagos Tortoises. Nomenclatural and survival status (Chelonian Research Foundation in association with Conservation International and Chelonia Institute)

Pritchard JA (1999) Preserving Yellowstone's natural conditions: science and the perception of nature. University of Nebraska Press, p. 370

Quiroga D (2009) Crafting nature: the Galápagos and the making and unmaking of a 'natural laboratory'. J Polit Ecol 16:123–140

Quiroga D (2013) Changing views of the Galapagos. In: Walsh SJ, Mena CF (eds) Science and conservation in the Galápagos Island: frameworks and perspectives. Springer, London, pp 23–48

Quiroga D, Rivas G (2016) Darwinian emergence, conservation, and restoration. Novel ecosystems and hybrid environments. In: Quiroga D, Sevilla A (eds) Darwin, Darwinism and conservation in the Galapagos Islands: the legacy of Darwin and its new applications. Springer, NY, pp 151–164

Reck G (2017) The Charles Darwin Foundation: some critical remarks about its history and trends. In: Quiroga D, Sevilla A (eds) Darwin, Darwinism and conservation in the Galapagos Islands: the legacy of Darwin and its new applications. Springer, NY, pp 109–134

Reynard E, Brilha J (eds) (2018) Geoheritage: assessment, protection and management. Elsevier, Amsterdam, 450 pp

Robinson E (1998) Tourism in geological landscapes. Geol Today 14 (4):151–153

Runte A (1979) National parks: the american experience. University of Nebraska Press, Lincoln

Sá A, Silva E, Rosado-González EM, Melo P, Palacio-Prieto JL (2017) Contribution for the discussion and new approaches about the development of UNESCO Global Geoparks in Latin America and the Caribbean. In: Almeida-Lima E, Nunes J-C, Meirinho P, Machado M (eds) European geoparks 14th conference, 7th–9th September 2017, Ponta Delgada, Azores, Pt: Abstracts Book, p 143

Sandelin B, Trautwein HM, Wundrak R (2014) A short history of economic thought. Routledge

Santander T, Gonzalez J, Tapia W, Araujo E, Montes C (2009) Tendencias de la investigación científica en Galapagos y sus implicaciones para el manejo del archipiélago en Tapia W., P Ospina, D Quiroga, J Gonzalez y C Montes eds. Ciencia para la Conservacion en Galapagos. PNG 2009

Sevilla A (2016) On the origin of species and the Galapagos Islands. In: Quiroga D, Sevilla A (eds) Darwin, Darwinism and conservation in the Galapagos Islands: the legacy of Darwin and its new applications. Springer, NY, pp 19–22

Spence MD (1999) Dispossessing the wilderness: Indian removal and the making of the national parks. Oxford University Press

Steinheimer F (2014) Charles Darwin's bird collection and ornithological knowledge during the voyage of H.M.S. "Beagle", 1831–1836. J Ornithol 145:300–320

Sulloway FJ (1984) Darwin and the Galapagos. Biol J Linn Soc 21(1–2):29–59

Trota AN, Pereira MJ (2015) Natural history of the Azores. Regional Directorate for the Environment (Azores Government), 139 pp

Valdivia G, Wolford W, Lu F (2014) Border crossings: new geographies of protection and production in the Galápagos Islands. J Ann Assoc Am Geogr 104(3)

Van de Haar E (2015) Degrees of freedom: liberal political philosophy and ideology. Transaction Publisher, New Brunswick and London

van Heijnsbergen P (1997) International legal protection of wild fauna and flora. IOS Press, Washington, DC. Chapter 2: "Historical survey" (§2.1–2.2), pp 9–17

Vesey-Fitzgerald DF (1969) National Park the best American idea. Wright Mendez in Yosemite started to do research on the population found in the parks. George Melendez Wright. San Francisco Family Salvado dinasties. Catlin idea first painter. John Muir Yosemite was exploited Roosevelt visit. 1916 park service

Villacis B, Carrillo D (2013) The Socioeconomic paradox of Galapagos. In: Walsh SJ, Mena CF (eds) Science and conservation in the Galápagos Island: frameworks and perspectives. Springer, London, pp 69–86

Walsh SJ, Mena CF (2013) Perspectives for the study of the Galapagos Islands: complex systems and human-environment interactions. In: Walsh SJ, Mena CF (eds) Science and conservation in the Galápagos Island: frameworks and perspectives. Springer, London, pp 49–68

Wikelski M, Foufopoulos J, Vargas H, Snell H (2004) Galápagos birds and diseases: invasive pathogens as threats for Island species. Ecol Soc 9(1) Art. 5

Wilson MC (2005) The "Placing" of identity in Nomadic Societies: aboriginal landscapes of the Northwestern Plains of North America. Mater History Rev 62(Fall 2005)/Revue d'histoire de la culture matérielle 62

Xun Z, Milly W (2002) National geoparks initiated in China: putting geoscience in the service of society. Episodes 25:33–37

Yang G, Chen Z, Tian M, Wu F, Wray RAL, Ping Y (2011) On the growth of national geoparks in China: distribution, interpretation, and regional comparison. Episodes 34(3):157–176

Zouros N, McKeever P (2008) European geoparks: tools for earth heritage protection and sustainable local development. In: Zouros N (ed) European geoparks. Natural History Museum of Lesvos Petrified Forest, Greece

5 The Geodiversity and Geoheritage of the Galapagos Islands: A Geotouristic Guide

5.1 Introduction

The exceptional geodiversity of the Galapagos Islands is available for all visitors, though the number of accessible sites is limited by the nature of the protections that are inherent within National Park areas. When planning a visit to the Galapagos Islands, therefore, in order to experience the geoheritage of this "cathedral of the natural sciences", there are logistical considerations to take into account and careful planning is essential.

The Galapagos National Park (GNP) manages 97% of the islands, and as such most of the sites that provide access to the geoheritage are on this land and managed by the park service. To help manage potential visitor impacts, the National Park authorities have identified a range of sites across the islands which can be visited subject to a range of conditions, depending on the factors such as the sensitivity of the site and its perceived carrying-capacity—hence quotas are usually in place. Most, however, will require the presence of a licensed guide, who typically arrange logistics, but must first be hired.

A list of designated tourist sites within the National Park provided on the GNP website (http://www.galapagos.gob.ec/), and reproduced here as Tables 5.1 for terrestrial sites and 5.2 for marine sites. Most of these sites are visited primarily for viewing marine or terrestrial flora or fauna—usually both—and most will also, inevitably, include geological features. It is important to emphasize that the lists in Tables 5.1 and 5.2 comprehensively list of *all* available tourist sites in Galapagos as designated by the GNP up to 2018. Very occasionally 'new' sites may be added to the lists, but very little change is envisaged in the near future.

As discussed in Chap. 2, in many other regions, states, or countries across the world, some form of inventory system or survey will have been applied to the area, as a basis for selecting key *geosites* for conservation as a *geoheritage*. However, in the Galapagos Islands, no systematic survey has yet been completed and due to the high level of protection in place, the only criteria it is currently possible to apply in this context is 'accessibility', to list those sites that are available for visit. That is not to say that there are not many other features of key geological interest throughout the islands, they are simply unavailable to visitors, with the rare exception of a permit granted by the GNP for those conducting scientific research. However, without a systematic survey, even this can become a self-repeating scenario, as prior knowledge of the location of such features is often essential before any application can be made—hence 'new' sites are likely to remain undocumented and unstudied. Outside of the National Park, of course, there will be many other available sites but again, without a systematic survey, it is currently not possible to provide a comprehensive list of those available to visit.

Table 5.3 lists a subset of the approved tour sites that are sites of particular geologic significance, and hence could be considered to be *geosites*. Due to the dominantly natural condition of the Galapagos Islands, most other sites, particularly those that are land-based, will offer some insight into the geologic activity, features, or history of the islands. However, we consider those in Table 5.3 to be the most ideal for experiencing and learning about the geologic features and processes that formed—and continue to form—the Galapagos Islands (as described in Chap. 2).

In addition to those sites selected from the GNP list, some additional geoheritage sites that are on either privately owned land or land that is owned and managed by the local municipalities on each island are also included in Table 5.3. As indicated above, this is not yet a comprehensive list, but it does cover most of the more spectacular lava tunnels that are open to tourists on private or publically-owned land. These sites will not be found on the list of GNP, but are nevertheless very important to include in any discussion of the geological heritage of the Galapagos, in particular as they are accessible and many are regularly visited. Table 5.3 also includes quarry sites on the islands of San Cristobal and Isabela that are adjacent to the municipalities of Puerto Baquerizo Moreno and Puerto Villamil, respectively. These sites not only demonstrate the internal structure of cinder cones, due to the way in which they have been excavated,

D. Kelley et al., *In the Footsteps of Darwin: Geoheritage, Geotourism and Conservation in the Galapagos Islands*, Geoheritage, Geoparks and Geotourism, https://doi.org/10.1007/978-3-030-05915-6_5

Table 5.1 List of all land based tourist sites designated by the Galapagos National Park

Island	Site	Activities
Bartolome	Bartolome	Walk
Daphne Mayor	Daphne Mayor	Walk
Espanola	Bahia Gardner	Walk
Espanola	Punta Suarez	Walk
Fernandina	Punta Espinoza	Walk
Floreana	Asilo de la Paz	Walk
Floreana	Bahia Post Office	Walk
Floreana	Cerro Allieri	Walk
Floreana	La Loberia	Walk
Floreana	La Misionera	Walk
Floreana	Mirador de la Baronesa	Walk
Floreana	Punta Cormorant	Walk
Genovesa	Bahia Darwin	Walk
Genovesa	El Barranco	Walk
Isabela	Bahia Cartago Grande	Walk
Isabela	Bahia Urbina	Walk
Isabela	Calera	Walk
Isabela	Caleta Tagus	Walk
Isabela	Centro de Crianza Arnaldo Tupiza	Walk
Isabela	Complejo de Humedales	Walk
Isabela	Concha de Perla	Walk
Isabela	Cueva de Sucre	Walk
Isabela	La Loberia	Walk
Isabela	Los Tuneles	Walk
Isabela	Minas de Azufre	Camp
Isabela	Mirador del Mango	Walk
Isabela	Muro de las Lagrimas	Walk
Isabela	Punta Albemarle	Walk
Isabela	Punta Moreno	Walk
Isabela	Tintoreras	Walk
Isabela	Volcan Alcedo	Walk
Isabela	Volcan Sierra Negra/Volcan Chico	Camp
Isabela	Loberia Grande II	Walk
Isabela	Punta Tortuga Negra	Walk
Mosquera	Mosquera Norte	Walk
Plazas	Plaza Sur	Walk
Rabida	Rabida	Walk
San Cristobal	Bahia Rosa Blanca	Walk
San Cristobal	Bahia Sardina	Walk
San Cristobal	Centro de Crianza Jacinto Gordillo	Walk
San Cristobal	Centro de Interpretacion	Walk
San Cristobal	Cerro Brujo	Walk
San Cristobal	Cerro Tijeretas	Walk
San Cristobal	El Junco	Walk

(continued)

Table 5.1 (continued)

Island	Site	Activities
San Cristobal	Galapaguera Natural	Walk
San Cristobal	Isla Lobos	Walk
San Cristobal	Jardin de las Opuntias	Walk
San Cristobal	La Loberia	Walk
San Cristobal	Manglecito	Camp
San Cristobal	Playa Baquerizo	Walk
San Cristobal	Playa del Muerto	Walk
San Cristobal	Playa Ochoa	Walk
San Cristobal	Puerto Chino	Camp
San Cristobal	Puerto Grande	Camp
San Cristobal	Punta Carola	Walk
San Cristobal	Punta Pitt	Walk
San Cristobal	Punta Pucuna	Walk
San Cristobal	La Tortuga	Walk
Santa Cruz	Bahia Ballena	Walk
Santa Cruz	Bahia Borrero	Walk
Santa Cruz	Centro de Crianza Fausto Llerena	Walk
Santa Cruz	Cerro Dragon	Walk
Santa Cruz	Garrapatero	Camp
Santa Cruz	Isla Eden	Walk
Santa Cruz	La Fe	Walk
Santa Cruz	Laguna de Las Ninfas	Walk
Santa Cruz	La Ratonera	Walk
Santa Cruz	Las Grietas	Walk
Santa Cruz	Las Palmitas	Walk
Santa Cruz	Los Gemelos	Walk
Santa Cruz	Mirador de los Tuneles	Walk
Santa Cruz	Playa Estacion Charles Darwin	Walk
Santa Cruz	Playa de los Perros	Walk
Santa Cruz	Playa Las Bachas	Walk
Santa Cruz	Tortuga Bay	Walk
Santa Cruz	Puntudo Cerro Crocker	Walk
Santa Cruz	Reserva El Chato	Walk
Santa Cruz	Playa Escondida	Walk
Santa Fe	Santa Fe/Fondeadero	Walk
Santiago	Bahia Sullivan	Walk
Santiago	Minas de Sal	Walk
Santiago	Playa Espumilla	Walk
Santiago	Puerto Egas	Walk
Santiago	Sombrero Chino	Walk
Seymour Norte	Seymour Norte	Walk

Most are available for walking tours, while a few allow camping

Table 5.2 All tour sites designated by the Galapagos National Park that are accessible by water only

Island	Site	Activities
Baltra	Baltra Noreste	Sn, Dg, Dv
Bartolome	Roca Felipe	Sn, Dv
Bartolome	Bartolome Punta	Sn, Dv
Daphne Menor	Daphne Menor	Ky, Dv
Darwin	El arco	Dg, Dv
Darwin	El Arenal	Dg, Dv
Espanola	Bajo Gardner	Sn
Espanola	Isla Gardner	Sn, Dg
Espanola	Islote Osborn	Sn
Fernandina	Cabo Douglas	Dv
Fernandina	Cabo Hammond	Dv
Fernandina	Punta Mangle	Sn, Dg
Floreana	Corona del Diablo	Sn
Floreana	Islote Caldwell	Dv
Floreana	Islote Champion	Sn, Dg, Dv
Floreana	Islote Enderby	Dv
Floreana	Islote Gardner	Dv
Floreana	Islote Watson	Dv
Floreana	La Botella	Dv
Isabela	Bahia Elizabeth	Dg, Ky
Isabela	Cabo Marshall	Sn, Dg, Dv, DvN
Isabela	Ciudad de las Mantas	Sn, Dg, Dv
Isabela	El Condenso	Sf
Isabela	El Estero	Sf
Isabela	El Finado	Sn, Dg, Fi
Isabela	Isla Cowley	Sn, Dv
Isabela	Islote Tortuga	Sn, Dg, Dv, Fi
Isabela	Islote Tortuga Oeste	Dv
Isabela	La Angelita	Sf
Isabela	Playa Grande	Sf
Isabela	Puerto Coca	Dv
Isabela	Punta Viente Roca	Sn, Dg, Dv
Isabela	Roca Blanca	Sn, Dg, Dv, Fi
Isabela	Roca Redonda	Sn, Dv
Isabela	Loberia Chica	DvI
Isabela	Arenero de El Faro	DvI
Isabela	Islote Cuatro Hermanos	Sn, Dg, Dv
Marchena	Playa Negra	Sn, Dg
Marchena	Punta Mejia	Sn, Dg, Dv
Marchena	Punta Espejo	Dv
Marchena	Punta Montalvo	Sn, Dg, Dv
Pinta	Cabo Chalmers	Dv
Pinta	Cabo Ibbetson	Dv
Pinta	Puerto Posada	Dv
Pinta	Punta Nerus	Dv
Pinzon	Bahia Pinguino	Sn, Dg, Dv

(continued)

Table 5.2 (continued)

Island	Site	Activities
Pinzon	Islote Dumb	Sn, Dv
Pinzon	Roca Sin nombre	Dv
Plazas	Plaza Norte	Sn, Dg, Ky, Dv
San Cristobal	Bajo de Cerro Brujo	Dv
San Cristobal	Caragua	Dv, DvI
San Cristobal	El Canon	Sf
San Cristobal	Five Fingers	Dv
San Cristobal	Islote Punta Pitt	Sn, Dg, Dv
San Cristobal	La Predial	Dv, DvN, DvI
San Cristobal	Leon Dormido	Sn, Dv, Sf
San Cristobal	Lolo Surf	Sf
San Cristobal	Los Crateres	Sf
San Cristobal	Pared Tijeretas	Dv, DvI
San Cristobal	Roca Ballena	Dv
San Cristobal	Tonga Reef	Sf
San Cristobal	Outer Reef	Sf
Santa Cruz	Bajo Bazan	Sn, Dg, Dv, DvI
Santa Cruz	Caamano	Dg, Ky
Santa Cruz	Caleta Trotuga Negra	Sf
Santa Cruz	Cerro Gallina	Dv, DvI
Santa Cruz	El Bajo	Dv, DvI, DvN
Santa Cruz	El Barranco	Sn, Ky, Dv
Santa Cruz	Guy Fawkes Sur	Dn, Ky, Dv
Santa Cruz	Piedra Ahogada	Sn, Dg, Fi
Santa Cruz	Punta Angermeyer	Sf
Santa Cruz	Punta Carrion	Sn, Dg, Dv
Santa Cruz	Punta Carrion Exterior	Dv
Santa Cruz	Punta Estrada	Ky, Dv, Sf, DvN, DvI
Santa Cruz	Punta Oeste Tortuga Bay	Sf
Santa Cruz	Rocas Gordon	Sn, Dg, Dv
Santa Fe	La Botella	Sn, Dg, Fi
Santa Fe	Punta del Miedo	Dv
Santa Fe	La Encanada	Sn, Dv
Santiago	Caleta Bucanero	Sn, Dg, Ky
Santiago	Islote Albany	Sn, Dg, Dv
Santiago	Roca Cousins	Sn, Dg, Dv
Santiago	Rocas Bainbridge	Dg, Dv
Santiago	Rocas Beagle	Sn, Ky, Dv
Seymour Norte	Seymour canal	Sn, Dv
Seymour Norte	Seymour Noreste	Sn, Dv
Wolf	Anchor Bay	Dg, Dv
Wolf	El Derrumbe	Sn, Dg, Dv
Wolf	Islote la Ventana	Dg, Dv
Wolf	La Banana	Dg, Dv
Wolf	Punta Shark Bay	Sn, Dg, Dv

Third column indicates activities that are available at each site: Snorkel (Sn), Dinghy Ride (Dg), Kayak (Ky), Diving (Dv), Surfing (Sf), Fishing (Fi), Night Diving (DvN), Diving Instruction (DvI) but also demonstrate a connection between geodiversity and the human societal activity on the islands. Section 5.2 contains further description and photographs of each of these geosites, and Table 5.3 lists them by island.

As indicated above, whatever the primary purposes for visiting the Galapagos Islands, be it for research, education, or tourism, there are logistical considerations to be considered that are specific to the Islands. Due to the Special Law of Galapagos of 1996, there are numerous restrictions in place that range beyond the protection of land and marine areas to cover aspects such as hotels, tour ships, and even the number of flights into the islands. Hence, unlike most other places on Earth, the Galapagos are not a place where visitors are able to arrive with no plans, rent a car and explore as they wish, hence some prior planning is necessary. When planning a visit to the islands, there are generally two options; a tour of the islands by boat or land-based tours, typically linked by an element of 'island-hopping'. Each option is introduced below, followed by an island-by-island guide to the notable—and accessible—geosites (Fig. 5.1).

With any type of visit to the Galapagos Islands, however, registered Naturalist Guides licensed by the Galapagos National Park will play an important role. There are over 800 licensed guides across the islands, and most are native to the islands. To obtain this title, the guides must participate in a training program that includes education on the ecology and geology of the islands, the laws permitting tourism, and the rules and regulations of the National Park. These guides participate in regular professional development activities in which they are trained further on the current state of scientific knowledge and newly permitted visitor sites. As indicated above, for visits, within National Park areas, the presence of a guide may be obligatory.

5.1.1 Tour by Marine Vessel

One option when visiting Galapagos is to arrange a visit that is sea-based. In this case, passengers stay on their cruise ship each night, and travel from island to island following a predetermined itinerary.

All boats used for such tours must be authorized by the Galapagos National Park for the transport of passengers who can spend the night on board and tour the authorized visits sites of the entire archipelago according to assigned itineraries that are specifically approved for each boat. Normally each boat has two approved itineraries, one consisting of touring the center, west and north of Galapagos, and the other consisting of the central, eastern and southern Galapagos. The Galapagos National Park tries to ensure that each tourist boat is assigned fixed itineraries that are maintained throughout the year with a rotation system between the center—west route and the north of the Galapagos and the central route—east and south of the Galapagos (Fig. 5.2).

As the Galapagos is a small paradise of evolution with unique species, many of which can be sensitive to disturbance to large-scale tourist pressure, some changes in the detailed itinerary may occur depending on the carrying capacity of individual sites, as determined by the GNP.

While there are no designated closed seasons and tourism remains open throughout the year, the tourist activity is aimed to be of an ecological nature, oriented towards the enjoyment of nature in a responsible manner. Aspects of regulating this impact are incorporated within the Special Law of Galapagos, which was intended to guarantee the conservation and sustainable development of the islands. Among the controls put in place to limit environmental impacts, is the regulation of cruise ships to have a capacity of up to 16 passengers, respecting the right of those boats that were already operating to continue when the law was applied. However, in recent years, the number of tourist boats of greater capacity have been increasing to such an extent that today there are now boats with capacities of 20, 32, 48, 90 and 100 passengers operating around the islands.

While the itinerary on a vessel based tour is predetermined and often inflexible, the advantage of this method of visit is that the boats will usually visit more islands than the typical land-based tourist would see. Many islands, while uninhabited, have landing sites for short walking tours (see Sect. 5.2).

5.1.2 Land-Based Tours

The alternative option when visiting the Galapagos Islands is to create a land based tour. In this case, visitors will stay in hotels on the inhabited islands, spend much of their time on day tours on the islands or on the water, and will commute from island to island on ferry boats. It is important to hire the services of a GNP Specialized Guide to obtain good technical and logistical information which makes the visit to the islands much easier. Each authorized guide is able to accompany a group of up to 16 passengers and can provide their professional services for a full day or for a half day. Visitors are free to walk around and explore municipal areas as they please, but almost all areas outside of the municipalities are part of the National Park where accompaniment by a licensed guide is required. As the Galápagos are still a community where many people know each other, in any travel agency, hotel, taxi or even talking with a member of the community it is possible to locate a guide to hire for a half day or full day of tours (Fig. 5.3).

The Special Law of Galapagos also dictates some aspects of land-based activities, such as the size of hotels, and the number of beds that each is allowed. Hotels are typically

Table 5.3 Geosites in Galapagos arranged by island

Geosite	Class	Description
San Cristobal Island		
Frigatebird Hill	Boat Tour Walking Tour	Cinder cone dissected by wave action. Accessible by land or visible by boat
Punta Pucuna	Boat/Hiking	Small Cinder Cones and lava flows. Fresh and well exposed
Tuneles de Eva	Boat/Hiking	Small lava tunnels within the lava flows of Punta Pucuna
Cerro Brujo/Witch Hill	Boat tour Mapping	Tuff cone with dikes
Kicker Rock	Boat/Snorkeling	Tuff cone
Punta Pitt	Boat/Hiking	Olivine beach. Tuff cone
Cerro Quemado	Municipal Quarry	Active quarry with dissected cinder cone, dikes
Santa Cruz Island		
Los Gemelos	Hiking (short distance)	Two large collapse craters
Mirador Tunnel	Hiking (short distance)	Short section of lava tunnel with features very well exposed. Close to town. Very short hiking
Primicias Tunnel	Private	Long lava tunnel. Tortoise reserve
Tuneles del Amor	Private	Long lava tunnel
Rancho Chato 2	Private	Long lava tunnel. Branching laterally
Royal Palm Galapagos	Private	Lava tunnels with multiple levels vertically
Hueco Grande	Private	Large collapse crater
Cerro Mesa	Private	Eroded cinder cone. Lookout tower on top
Isabela Island		
Sierra Negra	Hiking	Walk to top and on trail around rim of large caldera. Striking views
Sierra Negra Sulfur Mine	Hiking (long distance)	Long hike to view fumaroles and sulfur deposits
Volcan Chico	Hiking (long distance) or horses	Parasitic cones and fresh lava features on the flank of Sierra Negra
Las Tintoreras	Boat/Hiking/Snorkeling	Short hike on basaltic lava flows
Tagus Cove	Boat/Hiking	Tuff cone with lake. Volcanic deposits of Darwin Volcano
Urbina Bay	Boat/Hiking	Uplifted coral deposits exposed above sea level. Rhyolite deposits. Flank of Alcedo Volcano
Cueva de Sucre	Hiking (short distance)	Branching lava tunnel with single point entrance/exit
Cerro Pelado	Municipal Quarry	Active Quarry with dissected cinder cone, cluster of cinder cones, and fresh lava flows. Near town
Airport	Municipal Area	Uplifted beach sands
Santiago Island		
Sullivan Bay	Boat/Hiking	Excellent exposure of basaltic lava flows and features
Sombrero Chino	Boat/Hiking	Spatter cone with many lava tubes. Pillow lavas
Escalera de Bartolome	Boat/Hiking	Spatter cone. Young basaltic lava features

This table provides a subset of the designated tourist sites of the Galapagos National Park that contain an aspect of geoheritage as well as private or municipal sites that are not part of the park but are also of interest with regard to geoheritage

small and locally owned, although this is beginning to change. With increasing number of visitors each year, it is important to book lodging well in advance of travel. When commuting around an island, taxis are used, which also licensed by the government, which also controls their number on each island, but they are generally very easy to find. Each of the inhabited islands also, of course, has restaurants, night club/bars, souvenir shops, pharmacies, coffee shops, and public spaces, and many other ways to enjoy its towns.

In addition to guided tours to GNP sites, there are also privately owned sites which can be visited, typically for a fee

Fig. 5.1 Raul Salazar working as a Galapagos Naturalist Guide. *Photo* D. F. Kelley

of course. Additionally, on San Cristobal and Santa Cruz there are interpretation centers. Santa Cruz, in particular, is home to the Charles Darwin Foundation which has an exhibition hall open to tourists as well as gardens with characteristic plants and displays relevant to the flora and fauna of the islands and its conservation management.

5.1.3 Some General Advice ...

Remarkably, once in the islands, it can be very difficult to buy any sort of even vaguely technical guide, even to the wildlife, even at the Charles Darwin Research Center. Touristic shops tend to stock a range of typical gift items, most of which are not produced on the islands—an interesting challenge for any sustainable development project perhaps—and with a very variable range of printed works. It is essential, therefore, that any guides, for instance to wildlife, are ordered and purchased before arriving in the islands—examples are listed here in the bibliography. Geological guides are non-existent, so do not forget to pack this one!

5.2 Island by Island Guide to Geosites

In this section, we provide an inventory and description of the key Geoheritage sites that can be visited on each of the three principle islands, San Cristobal, Santa Cruz, and Isabela. These descriptions can form an ideal basis for visitors to build an itinerary, or to plan an educational trip to the islands (see Kelley and Salazar 2017). Section 5.2.4 provides descriptions of geosites on the other islands, which typically can be visited on day trips booked out of one of the three principle islands, or they are commonly be included in sea-based tours. General maps of the key islands are also provided, to help plan visits.

Fig. 5.2 Marine tour boats gather at Puerto Egas, Isla Santiago. *Photo* K. N. Page

5.2.1 San Cristobal Island

The tour sites, and more specifically, the geosites of San Cristobal Island are generally divided into two categories: there are those that are in the highlands and accessible by road, while many of the rest, even those that are listed as terrestrial sites, are around the coast of the island and accessible only by boat. When planning an itinerary to see a number of the sites on this island, it is important to be aware that one boat cannot take you everywhere that you want to go, as different boats and the guides who operate them are licensed only for certain sites. For example, one boat can take you on a tour that includes a landing at Punta Pucuna but the same boat cannot land at the beach on the south side of Witch Hill (Fig. 5.4).

Geosites:

Frigatebird Hill (Cerro Tijeretas) (0° 53′ 21.34″S, 89° 36′ 28.45″W): This hill is an older cinder cone (based on the amount of vegetation and erosion) that provides good exposures of its internal structure on the west side where it has been dissected due to erosive wave action. The site can be visited by a short walking trail from the Interpretation Center or by boat in Frigatebird Bay (Bahia Tijeretas). This bay is the site of the first stop made by the Beagle when the anchor was dropped on September 16, 1835, and so the area in often referred to as Darwin Bay or Bahia Darwin (Figs. 5.5 and 5.6).

Punta Pucuna (0° 44′ 18.57″S, 89° 26′ 15″W): This site is accessed by a guided boat tour out of Puerto Baquerizo Moreno. From a beach landing, it is possible to walk on relatively young lava flows (less than 1,000 years) with nicely exposed pahoehoe flow texture, hornitos, small cinder cones, tumuli, and some small lava tunnels (Figs. 5.7, 5.8 and 5.9).

Witch Hill (Cerro Brujo) (0° 45′ 39.11″S, 89° 27′ 51.92″ W): This large, dissected tuff cone is located along the western coast of the island. It is accessible by guided boat tour out of Puerto Baquerizo Moreno. By dinghy ride along the walls it is possible to examine the tuff and the dikes that are cutting vertically through the cone. By beach landing at Playa de la Salinas on the south side of the cone, it is possible to walk directly up to a wall of exposed tuff that has

Fig. 5.3 Tour group getting into island transportation on Isabela. *Photo* D. F. Kelley

been revealed due to erosion. Here, the tuffaceous material as well as fragments of basalt and carbonate rock are visible (Figs. 5.10 and 5.11).

Kicker Rock (Leon Dormido) (0° 46′ 42.08″S, 89° 31′ 07.33″W): This tuff cone is off of the west coast of the island and so is visited only by boat. The striking structure rises out of the ocean with very steep sides resulting from erosion of the original cone on all sides. The site is very popular for snorkeling and diving and there are tour options that combine Kicker Rock and Punta Pucuna or a landing on the south side of Witch Hill (Fig. 5.12).

Punta Pitt (0° 42′ 46.72″S, 89° 14′ 49.43″W): Located on the far northeastern tip of the island, this point is home to the largest tuff cone on the island. This site can be reached by a guided boat tour that will take passengers on a full circumnavigation around San Cristobal Island. A beach landing is possible at Punta Pitt and hiking provides a closer look at the structure and texture of the tuff (Fig. 5.13).

El Junco Lagoon (0° 53′ 43.00″S, 89° 28′ 52.36″W): This lake in the highlands of the island sits atop the shield volcano that makes the southwestern portion. There are not many additional volcanic features to be seen here because this section of the island is old, rising out of the sea ~5 million years ago and has a thick covering of soil. However, the lagoon itself is the largest standing freshwater body in the Galapagos Islands (Fig. 5.14).

Cerro Quemado (0° 54′ 45.53″S, 89° 36′ 51.76″W): This site is a large, active quarry on the southern end of the island, outside of the town of Puerto Baquerizo Moreno. Here, a large cinder cone has been dissected through excavation of material to be used in roads and other construction purposes. The layered interior structure of the cone can be seen (Fig. 5.15).

Interpretation Center (Centro de Interpretacion Ambiental 'Gianni Arismend') (0° 53′ 36.75″S, 89° 36′ 33.59″W): This is an essential stop for any visitor to San Cristobal Island. The center offers a great deal of information about the nature, natural history, human history, and present reality of life in the Galapagos Islands. The latter is particularly revealing and provides a lot of information related to

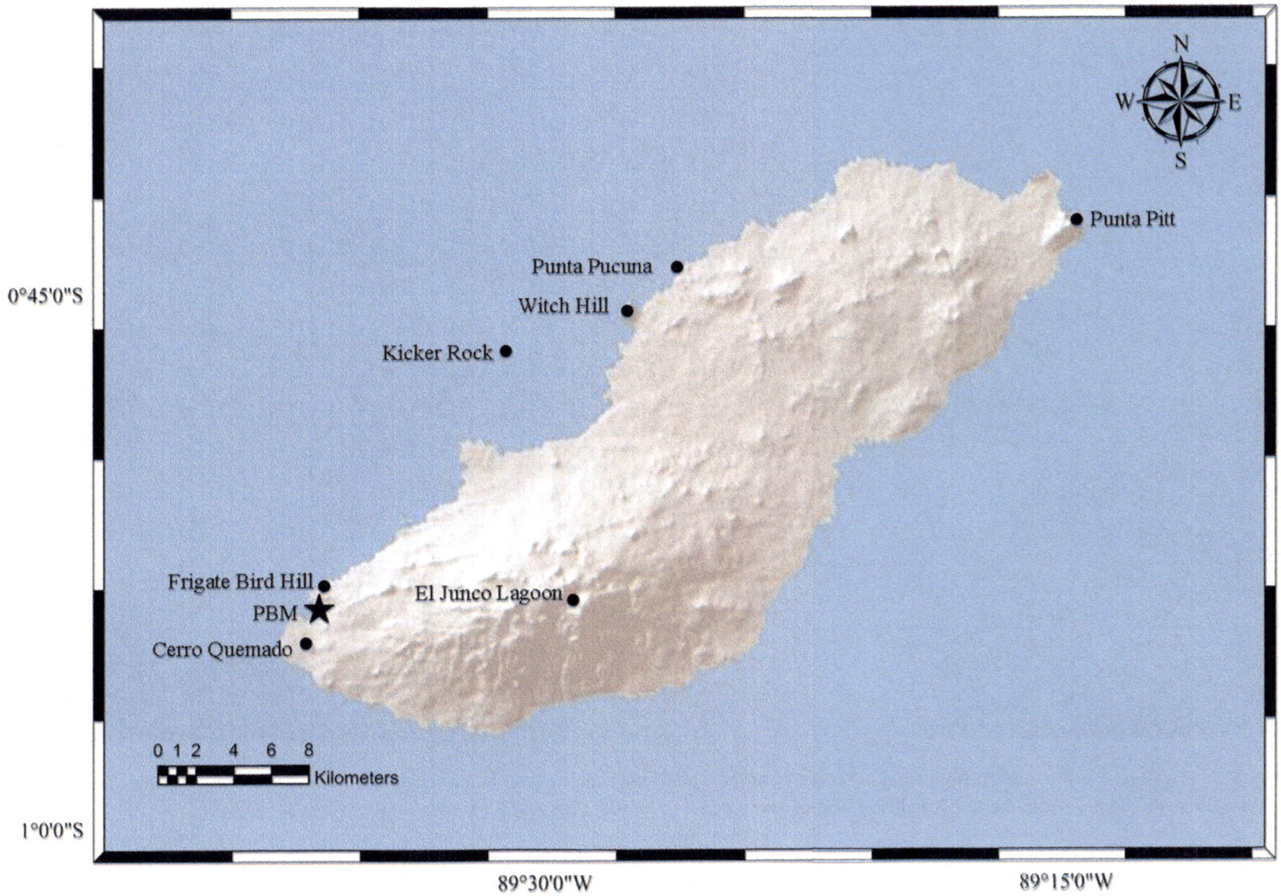

Fig. 5.4 Map of San Cristobal Island, Galapagos. Key geosites to be discussed in this section are located by black dots. The port town and governmental capitol of the Galapagos Islands, Puerto Baquerizo Moreno, is located with a black star

sustainable development projects and the reality of touristic exploitation of the islands today, with a large proportion of the profits still going offshore without any benefit to the resident population. This societal focus, contrasts with the conservation focus of the displays of the Charles Darwin Research Center on Santa Cruz, and hence two centers both contrast and complement each other. The Interpretation Center is located at the north end of the town of Puerto Baquerizo Moreno (Fig. 5.16).

5.2.2 Santa Cruz Island

Santa Cruz Island does not have any relatively young, fresh looking volcanic deposits or features. However, the island is large and much of it is accessible (unlike Isabela), owing to the main road that crosses it from Baltra and the airport in the north to the bustling city of Puerto Ayora on the south coast. The island has the largest resident population in the Galapagos, including many settlements in the highlands. This island also has the most visitors, again because of the airport and the largest city. For all of these reasons, there are plenty of sites for visitors. The main features are the collapse craters and the lava tunnels which are mainly in the highlands (Fig. 5.17).

Geosites:

Los Gemelos (0° 37′ 33.26″S, 90° 23′ 06.28″W): One of the most visited sites on Santa Cruz, these 'twin' collapse craters are situated on either side of the main road that comes across the island from the airport. The features are quite striking with the largest being up to 400 m in width and 700 m deep. Many layers of lava flows from the island's long history of eruption can be seen in the walls of the craters. The collapse craters formed as large lava tube systems in the subsurface collapsed, creating subsidence above (Fig. 5.18).

Mirador de los Tuneles (0° 43′ 55.41″S, 90° 19′ 42.15″W): This lava tunnel is a short distance outside of the city limits of Puerto Ayora along the main road that heads into the highlands. There is a small parking area just along

Fig. 5.5 Frigatebird Hill viewed from the bay. The flanks of the cinder cone can be seen on either side. The lower half of the cone contains more unconsolidated cinders, while the upper half is made of more massive basalt. *Photo* D. F. Kelley

Fig. 5.6 View from the top of Frigatebird Hill looking down on Darwin Bay with Kicker Rock, an eroded tuff cone, visible on the horizon to the right. *Photo* D. F. Kelley

Fig. 5.7 Tour group making a "wet landing" at Punta Pucuna. *Photo* D. F. Kelley

Fig. 5.8 Basaltic lava flow rocks of Punta Pucuna in the foreground, Witch Hill in the middle ground, and Kicker Rock in the distance. *Photo* D. F. Kelley

Fig. 5.9 Tour group walking on the lava flow features at Punta Pucuna. The landing beach is just out of the photo to the left. *Photo* D. F. Kelley

the road and then a short 100 m hike to the entrance of the tunnel. The tunnel extends a short distance back toward and under the road. The lava nicely displays flow textures along the walls and the floor indicating multiple phases of lava flowing through it. There is a skylight near the far end before the tunnel is blocked by boulders generated from roof collapse. While this is the shortest lava tunnel included here, it is the most accessible, being close to town (Fig. 5.19).

Primicias Tunnel (0° 40′ 09.92″S, 90° 25′ 48.40″W): This lava tunnel is accessible on a private property called Rancho Primicias and is also a well visited site due to it also being a tortoise reserve. A restaurant, coffee bar, and gift shop are added amenities. Visitors will often walk the property to see the many giant tortoises that reside there, and then head into the tunnel. The tunnel is a little over 400 m in length although in order to go in one end and out the other, there is a very low area that can only be passed by crawling (often in mud) towards the exit end. Visitors can opt to walk in from the entrance, turn around at the low spot and return back through the tunnel to the entrance. This low area is present because much of this tunnel has been filled partially by a subsequent lava flow. Therefore, much of the floor of the tunnel is the top of a lava flow that partially filled the tunnel and then cooled and solidified in place. Electric lights are provided along the length of the tunnel (Fig. 5.20).

Tuneles del Amor (0° 41′ 35.30″S, 90° 19′ 01.51″W): This lava tunnel is located on privately owned property in the highlands near the village of Bellavista. The tunnel is over 2 km long, but the portion that is available to visitors is around 1,000 m. This tunnel has electric lights along its entire length (Fig. 5.21).

Rancho Chato (20° 40′ 03.28″S, 90° 25′ 50.13″W): This is another privately owned site that is a popular tourist stop for the primary purpose of close up viewing of free roaming giant Galapagos tortoises. This property, like many in the highlands of Santa Cruz has a lava tunnel beneath the surface. This short, lighted lava tunnel can be incorporated into a stop at this destination (Fig. 5.22).

Royal Palm Galapagos (0° 38′ 55.98″S, 90° 23′ 35.99″W): On the privately owned property of this high end resort in the highlands (http://www.royalpalmgalapagos.com/), there are lava tunnels that are available for visit by tourists. The lava tunnels on this property are unique amongst the others on the island because they branch into tunnels that are stacked on top of one another vertically. The tunnels are well lit, have ladders and stairs, and visitors are required to wear lighted helmets that are provided.

Hueco Grande (0° 38′ 15.05″S, 90° 17′ 29.72″W): This site and the next, Cerro Mesa, are close together on privately owned land. There is a small restaurant/coffee bar where visitors can stop to pay a small token for the visit on the way in or out. Hueco Grande or "big hole" is a large collapse crater that is similar to those at Los Gemelos (Fig. 5.23).

Fig. 5.10 Witch Hill as viewed from dinghy ride. Here, two black colored dikes can be seen cutting vertically through the lighter colored tuff. *Photo* D. F. Kelley

Fig. 5.11 Witch Hill can be seen to be a cone with the edges eroded by wave action. *Photo* D. F. Kelley

Fig. 5.12 Kicker rock (Leon Dormido): **a** viewed as approached from the northwest; **b** a separation exists between this tall spire and the main body of Kicker Rock. Visitors are able to snorkel through crack from one side to the other looking for sea turtles, rays, and sharks; **c** a group of visitors snorkeling at Kicker Rock. A sea turtle (foreground) is joining. Massive and layered deposits of tuff are visible on the walls of the structure. The rising southern flank of San Cristobal Island is visible in the background

Fig. 5.13 Punta Pitt on the northwest corner of San Cristobal Island: **a** the nested tuff cones can be seen rising together. *Photo* D. F. Kelley; **b** looking closer, the layers of pyroclastic material (ash and Lapilli tephra) can be seen where the sea has eroded a vertical rock wall. *Photo* D. F. Kelley

Fig. 5.14 El Junco Lagoon. *Photo* D. F. Kelley

Fig. 5.15 Quarried cinder cone at Cerro Quemado. The outward sloping layers of loose scoria can be seen in the excavated wall of the quarry. *Photo* D. F. Kelley

Fig. 5.16 The Centro de Interpretacion Ambiental 'Gianni Arismend' at Pte. Baquerizo Moreno, Isla San Cristobal. *Photos* K. N. Page

Cerro Mesa (0° 38′ 33.55″S, 90° 17′ 06.23″W): Cero Mesa, or "Table Hill" is an old cinder cone that has built on the sloping flanks of the shield structure of the island of Santa Cruz. The cone is quite vegetated, so does not reveal many features. However, this is very typical of cinder cones across Santa Cruz, as due to the lack of recent volcanic activity on the island, all the islands cinder cones are vegetated. Cerro Mesa is worth a visit, however, because it is possible to drive to the top of the hill where an observation tower has been constructed. From the tower, it is possible to have a very good view of around one third of the island, and observe many islets and other coastal features, as well as other cinder cones, collapse craters, and depressions. This site and Hueco Grande belong to the same property which has a small restaurant/coffee bar.

Charles Darwin Research Station (0° 44′ 31.58″S, 90° 18′ 14.86″W): The Charles Darwin Foundation manages a multidisciplinary research station on Santa Cruz Island on the east side of Puerto Ayora (http://www.darwinfoundation.org/en/about-us/cdrs-visit/). Due to the pivotal role the Foundation has had in the development of the conservation practice across the islands and the GNP itself, this should be considered an essential stop for any visitor. The station maintains the Charles Darwin Exhibition Hall an informative visitor center, including panels, displays, and videos describing the scientific research efforts in the Galapagos Islands. Surrounding the center are gardens with examples of many famous endemic vegetation of the islands, and further information panels about species conservation. There is some geological content, but for socio-economic information about the islands, visiting the Interpretative Centre on San Cristobal should also be considered essential, if logistically possible (Fig. 5.24).

5.2.3 Isabela Island

Of the populated islands of Galapagos, Isabela is the furthest to the west, and so it is closest to the current volcanic hotspot. As a result, there is frequent eruptive activity from one

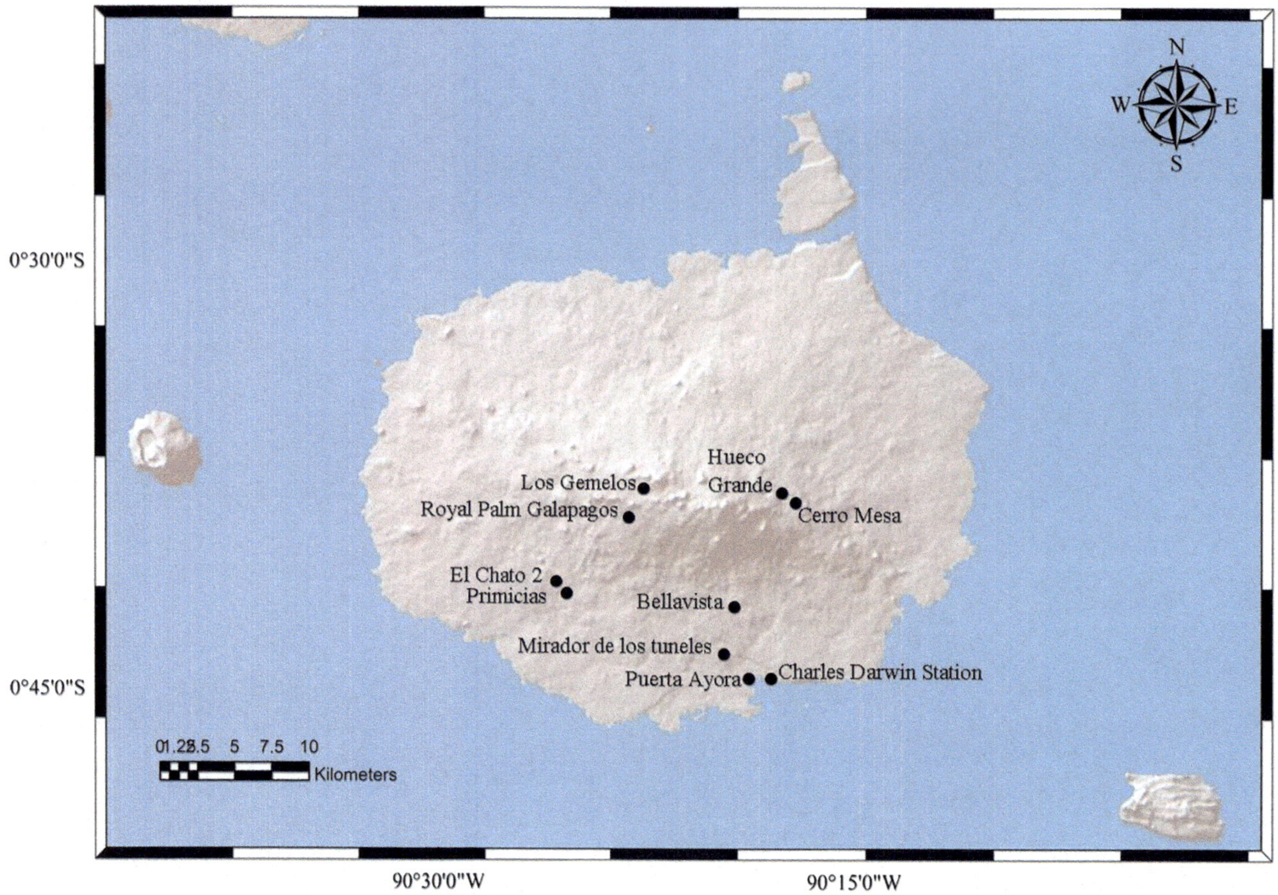

Fig. 5.17 Map of Santa Cruz Island

or other of the large shield volcanoes that connect to make this island, the most recent being on the west side of Sierra Negra volcano in 2018. The town of Puerto Villamil is on the distal flank of Sierra Negra, to the south-east. There are a number of geosites which can visited in and around the town of Puerto Villamil itself, while others are present close to the road which leads from the town up the flanks of the volcano to the caldera itself. Much of this largest island is inaccessible, however, as it lies within the National Park, but various tour sites have been designated around the coast of the island and are accessible by boat landing (Fig. 5.25).

Isabela is a two-hour boat trip from Puerto Ayora on Santa Cruz and some visitors travel to Isabela for a day trip in order to hike around the rim of Sierra Negra—but to experience more of the key geosites that this young, active island has to offer, it is best to stay for a few days. The small town of ~2000 residents has plenty of hotels and restaurants to accommodate visitors. The number of amenities has recently been growing every year with improvements also being made to the infrastructure of the town in order to attract more visitors.

Geosites:

Sierra Negra (0° 50′ 12.41″S, 91° 05′ 25.20″W): The volcano is most visited active volcano across the archipelago and can be accessed by road from Puerto Villamil, including by bus. The road ends close the top of the caldera at a parking area with restrooms. From there, visitors can either take a short but steep hike to the rim, or join an organized horse ride. It is possible hike or ride horses around the part of the rim of the large caldera at the top of this shield volcano where the views are spectacular. The stark black basaltic rock on the caldera floor, some of it representing flows from as recently as 2005 (Geist et al. 2008) contrasting sharply with the lush green vegetation on the walls and rim of the caldera (Fig. 5.26).

Sulfur Mine (0° 49′ 57.25″S, 91° 09′ 52.54″W): By hiking to the left, clockwise along the rim of Sierra Negra from the parking area, it is possible to arrive at the site of former Sulfur mines after a distance of 11 km (see Fig. 5.21). This is a longer hike than to reach Volcan Chico

Fig. 5.18 **a** One wall of the east crater of Los Gemelos showing the layers of lava flows that make up the upper crust on the flanks of these shield volcanoes. A vegetated cinder cone can be seen above in the near distance. *Photo* D. F. Kelley; **b** collapse crater on the west side of the road at Los Gemelos. *Photo* D. F. Kelley

Fig. 5.19 Mirador del Tuneles. Smooth flow texture can be seen on the lower wall, and "shelves" left higher up the wall where a roof of a lava flow started to form when the initial tunnel was reused by a later flow. *Photo* D. F. Kelley

in the other direction, but offers a visit to fumaroles and the sulfur deposits that have been made by mineral rich water running down the slope. Camping is permitted here (Fig. 5.27).

Volcan Chico (0° 46′ 32.18″S, 91° 06′ 06.34″W): The hike to this site from the starting point on the rim by the parking area is around 8 km and can be done on foot or horseback, with very little change in elevation. Volcan Chico site is a parasitic fissure vent on the flank of Sierra Negra which last erupted in 1979. The cones, flows, hornitos, and lava flow channels are still mostly unvegetated, and offer an excellent location to experience such features. On a clear day, this site also offers fantastic views of much of Isabela Island as well as Fernandina. A marked trail leads through the lava field and cinder cones (Figs. 5.28, 5.29 and 5.30).

Cueva de Sucre (0° 53′ 41.44″S, 91° 03′ 31.85″W): This site is located on the flanks of Sierra Negra volcano in the highlands between Puerto Villamil and the caldera rim. This "cave" is a group of branching lava tunnels and is referred to as such as there is only one entrance. This site is different from the lava tunnels of Santa Cruz because it has numerous branches that split laterally and come back together. It is possible to enter at the cave mouth, walk in and out of the different branches and exit again at the mouth. The entrance to the cave is visually striking with heavy vegetation surrounding it. The diameter of the tunnels is generally smaller than those in Santa Cruz. The ceiling of the tunnels in much of the cave is golden in color and shimmering. This is due to a combination of yellowish sulfur forming a thin layer on the basalt, and drops of water condensing on the surface to provide the luster. No lights are provided in the tunnels at this site (Figs. 5.31 and 5.32).

Cerro Pelado (0° 56′ 18.10″S, 90° 59′ 35.83″W): This area is not part of the National Park, and is owned by the municipality of Puerto Villamil. It is the site of a long, ridge-like cinder cone and its associated lava flow, associated with an area with around 12 nested, small cinder cones. These features were created by an eruption of only a few

Fig. 5.20 Visitors in a relatively low clearing in a portion of the lava tunnel at the Primicias Ranch in the highlands of Santa Cruz Island. *Photo* D. F. Kelley

hundred years ago. The site is now an active quarry worked by the municipality, excavating the cinder materials from the main cinder ridge. The lava flows fields around the site are quite treacherous for walking, having very jagged aa flow surfaces with blocks that are several meters across.

This relatively young lava flow is conspicuous for its location on the distal flank of the volcano just outside of Puerto Villamil, as the surrounding terrain is more weathered and vegetated.

This lava itself is much darker and black in color, mostly lacks vegetation and stands with a flow thickness of ~4 m above the surrounding surface. This site is interesting because it shows a possibility of eruption from a site other that the top of the volcano, which in this case was quite close to the current center of population. This is not a typical tourist site because of the mining activity, but it is a good stop to make with a group of geology students (Figs. 5.33 and 5.34).

El Chapin (0° 56′ 44.25″S, 90° 58′ 29.85″W): This site is a former municipal quarry on the edge of the city of Puerto Villamil, that now contains a brackish water lagoon, due to mixing of salt water and fresh water. As a result, it is a popular feeding site for flamingos (species?), and in turn a popular site for tourists. This quarry worked a cinder cone until suitable material was exhausted, when operations moved to Cerro Pelado. The original geologic landform is no longer present, save for some tephra layers that are visible on the wall of the quarry, but the human use of and effects on this geoheritage is interesting - and the close proximity to the town makes the site an easy one to include in most visits to Isabela (Fig. 5.35).

Puerto Villamil Airport (0° 56′ 42.89″S, 90° 57′ 17.41″W): Adjacent to the airport outside of Puerto Villamil, is a small area that has been quarried for calcareous sand. This is also not a GNP site, but rather is owned by the municipality. Quarrying has left a roughly 2-m-high outcrop of beach deposits,

Fig. 5.21 Lava tunnel at the "Tunnels of love" in Buena Vista. This unique tilted cross section can be observed at several places along the length of the tunnel where the direction of the tunnel takes a turn (to the right in this case). *Photo* D. F. Kelley

elevated due to changes in relative sea level and due to growth of the island. As one of the very few sites across the islands with sedimentary deposits, this is an important geosite and provides insight into the recent evolution of the islands and even shows interlayering of beach deposits and lava flows (Fig. 5.36).

Cabo Rosa (1° 02′ 56.21″S, 91° 10′ 08.08″W): This is a GNP marine site that is visited by boat on a day tour from Puerto Villamil. The tour includes a short hike on land to see the nesting sites of blue-footed boobies as well as snorkeling at two sites in the area. This is also an important geosite because it contains a unique geologic feature—many basaltic arches. These arches are the remnants of the roof of a cooling lava flow that entered the sea having come from the summit of Sierra Negra. As the lava flow settled into a depression in this flat coastal area, many basaltic columns formed between the solidifying roof of the lava flow and the floor. After the liquid lava had emptied out of the resulting chamber, the columns and roof were left behind to subsequently be eroded by wave action. What remains now are many thin arches of basalt spanning from one column to

Fig. 5.22 Development of tourist access to a Lava Tunnel on the El Chato tortoise ranch, Isla Santa Cruz (2016). *Photo* K. N. Page

another above the water level, whilst the columns can be seen underwater buy snorkeling (Fig. 5.37).

Isla Tortuga (1° 01′ 19.15″S, 90° 52′ 59.95″W): This site off of the coast of Isabela is another GNP marine site that is visited by day trip from Puerto Villamil. This site offers snorkeling and diving. The island is a "C" Shaped tuff ring that has been eroded on the southern side by the prevalent wave and wind action. While visits to this site do not include land-based activities, it provides one of the best opportunity to view tuffaceous material around Isabela Island (Fig. 5.38).

Urbina Bay (0° 23′ 32.36″S, 91° 13′ 48.72″W): This site is on the western side of Isabela Island at the base of Alcedo Volcano (See Fig. 5.21). The site is accessed by boat, and is mostly used by ship-based visitors. There is a hiking loop onto the island leading to the 'fossil beach', which comprises carbonate rocks formed on a beach and in the shallow water which includes coral structures. In 1954 these deposits and the living reef were rapidly uplifted to their current situation at 5 m above sea level, during which the coastline moved outwards by around ¾ mile.

Punto Moreno (0° 41′ 22″S, 91° 15′ 16″W): A relatively recently opened up GNP site on the south-western side of Isabella which is only accessible by boat. The visit includes a short walk across a spectacular and vast coastal lava platform of black basalt showing classic pahoehoe 'ropy lava' surfaces and other structures such as tumuli. The view of Sierra Negra volcano across the lava plain is one on the best across the island and shows its low domed shape 'shield volcano' shape perfectly (Figs. 5.39 and 5.40).

5.2.4 Sites on Other Islands

Visits to most of the other islands of the Galapagos are virtually only possible from boat-based tours, due to their distance from the main settlements of the inhabited islands, although some may still be accessible as day trips. Some areas include only marine activities such as diving or snorkeling, but a few also include landings during which the geology can be examined rather than just viewed from the water. Most of these sites will include geological features, but only the most notable locations are listed here.

Chinese Hat (0° 22′ 10″S, 90° 34′ 58″W): This small island is a cinder cone with relative young lava flows, including tunnels. Visits to the island by boat include a short

Fig. 5.23 Hueco Grande is a collapse crater that, like Los Gemelos, exposes many layered lava flows visible in the walls. *Photo* D. F. Kelley

coastal walk during which some of the lavas can be viewed (Fig. 5.41).

Isla Rabida (0° 3′ 59″S, 90° 42′ 39″W): The small island, just south of Santiago, is a scoria cone, erosion of which is responsible for producing its noted red beach where visitors come ashore by boat. The red color has been produced by the tropical weathering of the iron minerals with the basaltic scoria. Hiking trails on the island also pass a coastal brackish lagoon system (Fig. 5.42).

Fernandina: Punta Espinosa (0° 15′ 49″S, 91° 26′ 38″W): Although noted for its large marine iguana colony, Punta Espinosa also shows a range of basaltic lava features including pahoehoe surfaces and tumuli. The site is reached by boat. Fernandina is the youngest of the Galapagos islands (Fig. 5.43).

Fig. 5.24 The Charles Darwin Research Centre and botanical gardens, Puerto Ayora, Santa Cruz. *Photos* K. N. Page

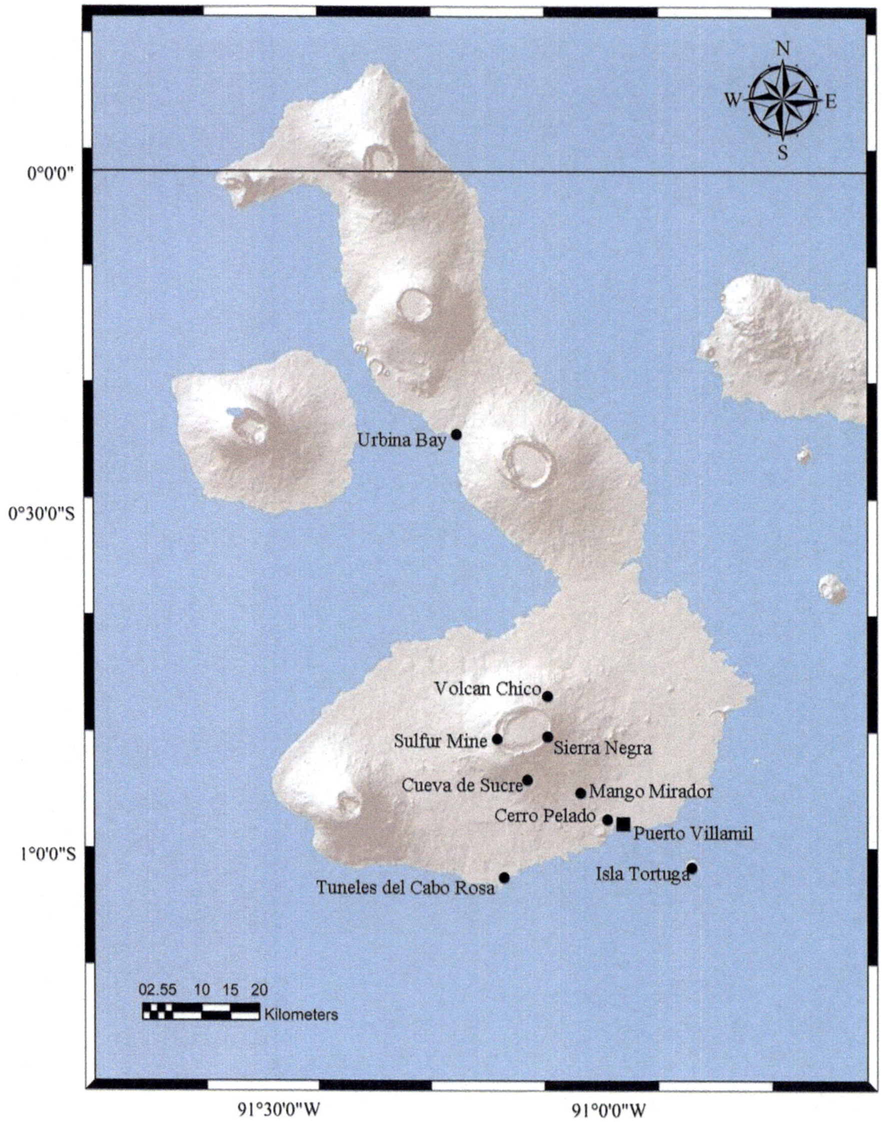

Fig. 5.25 Map of Isabela Island. The shield volcanoes on the island, from south to north, are Cerro Azul, Sierra Negra, Alcedo, Darwin, Wolf, and Ecuador. Fernandina Island is on the west side and Santiago Island is on the east

Santiago: Puerto Egas (0° 14′ 31″S, 90° 51′ 33″W): The area includes a tuff volcano with a lagoon with seawater and salt formation. The landing site, by boat, includes a cliff in fine-grained bedded ash deposits and a hiking trail leads to a shore platform in black basalt with well-developed pahoehoe structures and probable tumuli. A well-known blow hole in this surface is known as "Darwin's Toilet" because it fills and empties with water from below. This and other channels in the lava surface suggest the presence of collapsed lava tunnels (Figs. 5.44, 5.45, 5.46, 5.47 and 2.30b).

Fig. 5.26 Sierra Negra Volcano: **a** Visitors hiking the rim of Sierra Negra Volcano. The interior of the caldera crater is to the left. The younger, unvegetated area around Volcan Chico can be seen on the right flank of the volcano having erupted in the last century. In the distance, the silhouette of Alcedo Volcano can be seen beneath the clouds. La Cumbre Volcano at the center of Fernandina Island can be seen faintly just above the hikers. *Photo* D. F. Kelley; **b** it is also an option to view the caldera via horseback riding along the rim. *Photo* D. F. Kelley; **c** the black basaltic rock that is covering the caldera floor was emplaced by a lava flow during the eruption of 2005. The sulfur mine site can be seen on the far side of the caldera (hill with whitish color). In the far distance, Cerro Azul Volcano can be seen. *Photo* D. F. Kelley

Fig. 5.27 Sulfur mines from the inner slope of the caldera wall. The white mineral deposits are seen covering the slope beneath the fumaroles. *Photo* D. F. Kelley

Fig. 5.28 Volcan Chico area with cinder cones and lava flows. In the distance are Alcedo Volcano and Darwin Volcano. Ecuador Volcano is faintly visible above the cinder cone on the left of the photo. *Photo* D. F. Kelley

Fig. 5.29 The largest cinder cone at the Volcan Chico site. *Photo* D. F. Kelley

Fig. 5.30 Hiking trail through the Volcan Chico site. *Photo* D. F. Kelley

Fig. 5.31 Entrance to the Cueva de Sucre lava tunnels. *Photo* D. F. Kelley

Fig. 5.32 Interior of Cueva de Sucre showing the shiny gold coating on the ceiling of the cave. This tunnel height is 2–3 m. *Photo* D. F. Kelley

Fig. 5.33 Mining activity at Cerro Pelado. The coast line to the northeast is visible with the "four brothers" tuff cone islands near the shore. *Photo* D. F. Kelley

Fig. 5.34 Mining operation at Cerro Pelado. View from atop the cinder ridge. The black lava flow associated with this eruptive center can be seen to make a sharp contrast with the surrounding vegetated, older lava flows. *Photo* D. F. Kelley

Fig. 5.35 Flamingos feeding in El Chapin lagoon, with layers of basaltic pyroclastic material in the wall of the old quarry above. *Photo* D. F. Kelley

Fig. 5.36 Former beach deposits exposed by quarrying near Puerto Villamil Airport. *Photo* D. F. Kelley

Fig. 5.37 Los Tuneles de Cabo Rosa. These basalt arches span from one column to another just above the level of the water. *Photo* D. F. Kelley

Fig. 5.37 (continued)

Fig. 5.38 Flank of Isla Tortuga showing characteristic tan color of tuffaceous material where the island is not vegetated. *Photo* D. F. Kelley

Fig. 5.39 Arrival at Punta Moreno, with view of Sierra Negra beyond and blocky lava flow entering the sea at the landing site (Isla Isabela). *Photo* K. N. Page

Fig. 5.40 Lava features on the tourist trail from Punta Moreno: top left—pahoehoe flow surface; top right—collapse structure, now flooded; bottom left—tumuli; bottom right—cooling cracks on lave flow surface. *Photos* K. N. Page

Fig. 5.41 'Chinese Hat' (Sombrero Chino): eroded scoria cone (above—seen on approach by sea; below—from the tourist trail on the beach). *Photo* K. N. Page

Fig. 5.42 Red sand beach ('Playa Roya') and cliffs of eroded scoria on Isla Rabida. *Photo* K. N. Page

Fig. 5.43 Pahoehoe lava surface and tumuli at Punta Espinosa, both beside the tourist trail. *Photos* K. N. Page

Fig. 5.44 Landing beach at Puerto Egas showing fine-grained bedded tuff, probably water-lain, in the cliffs and eroded as arches in the headland. *Photo* K. N. Page

Fig. 5.45 Coastal platform in black basalt with well-developed pahoehoe structures on the tourist trail from Puerto Egas. *Photo* K. N. Page

Fig. 5.46 Cinder cone at Puerto Egas and beach platform showing worn pahoehoe flow structures, overlain by bedded tuff at back of shore area. *Photo* K. N. Page

Fig. 5.47 Probable 'squeeze-up' structures in lava flow on the tourist trail from Puerto Egas. *Photo* K. N. Page

References

Geist D et al (2008) The 2005 eruption of Sierra Negra Volcano, Galápagos, Ecuador. Bull Volcanol 70:655–673

Kelley DF, Salazar R (2017) Geosites in the galápagos islands used for geology education programs. Geoheritage. https://doi.org/10.1007/s12371-016-0190-3

Printed by Printforce, the Netherlands